ANNALS OF THE NEW YORK ACADEMY OF SCIENCES

Volume 577

EDITORIAL STAFF

Executive Editor
BILL BOLAND

Managing Editor
JUSTINE CULLINAN

Associate Editor
KEVIN M. MAYER

The New York Academy of Sciences
2 East 63rd Street
New York, New York 10021

THE NEW YORK ACADEMY OF SCIENCES
(Founded in 1817)

BOARD OF GOVERNORS, 1989

WILLIAM T. GOLDEN, *Chairman of the Board*
LEWIS THOMAS, *President*
CHARLES A. SANDERS, *President-Elect*

Honorary Life Governors

SERGE A. KORFF   H. CHRISTINE REILLY   IRVING J. SELIKOFF

Vice-Presidents

MARVIN L. GOLDBERGER   DAVID A. HAMBURG   CYRIL M. HARRIS
DENNIS D. KELLY   PETER D. LAX

HENRY A. LICHSTEIN, *Secretary-Treasurer*

Elected Governors-at-Large

GERALD D. LAUBACH   JOHN D. MACOMBER   NEAL E. MILLER
LLOYD N. MORRISETT   GERARD PIEL   JOSEPH F. TRAUB

FLEUR L. STRAND, *Honorary Past Chair*   HELENE L. KAPLAN, *General Counsel*

OAKES AMES, *Executive Director*

ETHICAL ISSUES ASSOCIATED WITH SCIENTIFIC AND TECHNOLOGICAL RESEARCH FOR THE MILITARY

ANNALS OF THE NEW YORK ACADEMY OF SCIENCES
Volume 577

ETHICAL ISSUES ASSOCIATED WITH SCIENTIFIC AND TECHNOLOGICAL RESEARCH FOR THE MILITARY

Edited by Carl Mitcham and Philip Siekevitz

The New York Academy of Sciences
New York, New York
1989

Copyright © 1989 by the New York Academy of Sciences. All rights reserved. Under the provisions of the United States Copyright Act of 1976, individual readers of the Annals *are permitted to make fair use of the material in them for teaching or research. Permission is granted to quote from the* Annals *provided that the customary acknowledgment is made of the source. Material in the* Annals *may be republished only by permission of the Academy. Address inquiries to the Executive Editor at the New York Academy of Sciences.*

Copying fees: *For each copy of an article made beyond the free copying permitted under Section 107 or 108 of the 1976 Copyright Act, a fee should be paid through the Copyright Clearance Center, 21 Congress Street, Salem, MA 01970. For articles of more than 3 pages, the copying fee is $1.75.*

♾ *The paper used in this publication meets the minimum requirements of American National Standard for Information Sciences—Permanence of Paper for Printed Library Materials, ANSI Z39.48-1984.*

The illustration on the cover of the paperback volume depicts Thetis receiving Achille's new armor from Hephaestus. Originally a fire god, Hephaestus later became a devine smith, fashioning ornaments, weapons, and magical contrivances for the gods and heroes.

Library of Congress Cataloging-in-Publication Data

Ethical issues associated with scientific and technological research for the military/edited by Carl Mitcham and Philip Siekevitz.
    p. cm.—(Annals of the New York Academy of Sciences, ISSN 0077-8923; v. 577)
    Includes bibliographical references.
    Result of a conference held Jan. 26-28, 1989, in Arlington, Va., sponsored by the New York Academy of Sciences and Polytechnic University.
    ISBN 0-89766-559-7 (alk. paper).—ISBN 0-89766-560-0 (pbk.: alk. paper)
    1. Military research—Moral and ethical aspects—Congresses.
2. Military research—United States—Moral and ethical aspects—Congresses. I. Mitcham, Carl. II. Siekevitz, Philip. III. New York Academy of Sciences. IV. Polytechnic University (Brooklyn, New York, N.Y.) V. Series.
Q11.N5 vol. 577
[U390]
500 s—dc20   89-13779
[174'.95]   CIP

CCP
Printed in the United States of America
ISBN 0-89766-559-7 (cloth)
ISBN 0-89766-560-0 (paper)
ISSN 0077-8923

ANNALS OF THE NEW YORK ACADEMY OF SCIENCES

Volume 577
December 29, 1989

ETHICAL ISSUES ASSOCIATED WITH SCIENTIFIC AND TECHNOLOGICAL RESEARCH FOR THE MILITARY[a]

Editors and Conference Organizers
CARL MITCHAM and PHILIP SIEKEVITZ

CONTENTS

Preface. By CARL MITCHAM and PHILIP SIEKEVITZ	ix
Foreword: The Genetic and Psychological Basis of Warfare as a Challenge to Scientific Research. By GEORGE BUGLIARELLO	xv

Part I. Main Themes

The Spectrum of Ethical Issues Associated with the Military Support of Science and Technology. By CARL MITCHAM	1
The Moral Arguments for Military Research. By ROBERT H. DINEGAR	10
The Moral Arguments against Military Research. By BERNARD ROTH	21
Masters of War: The Moral Arguments and the Traditions of Ethics. By DOUGLAS MACLEAN	34
The Moral Arguments and the Traditions of Religious Ethics. By ROGER L. SHINN	40
The Moral Arguments and the Philosophy of Science and Technology. By PAUL T. DURBIN	47

Part II. Historical and Conceptual Background

Hephaestus and History: Scientists, Engineers, and War in Western Experience. By ALEX ROLAND	51
Drawing the Line: An Examination of Conscientious Objection in Science. By ROSEMARY CHALK	61

[a] This volume is the result of a conference entitled **Ethical Issues Associated with Scientific Research for the Military**. The conference, which was sponsored by the New York Academy of Sciences and Polytechnic University, was held in Arlington, Virginia on January 26–28, 1989.

The Fine Structure of Military Research as an Essential Background for Discussions of Ethics: Remarks on a New Liberation Ethics. *By* RUSTUM ROY 75

Science Pure and Applied as Essentially Involved with Ethics. *By* KRISTIN S. SHRADER-FRECHETTE 86

Remarks on Ethics and Values Studies at the National Science Foundation. *By* RACHELLE D. HOLLANDER 94

Part III. Special Issues and Debates

The Just War Ethic and the Military Support of Science and Technology. *By* J. BRYAN HEHIR 99

The Social/Professional Responsibility of Engineers. *By* DEBORAH G. JOHNSON 106

Conducting Scientific Research for the Military as a Civic Duty. *By* KENNETH W. KEMP 115

Military Funds, Moral Demands: Personal Responsibilities of the Individual Scientist. *By* DOUGLAS P. LACKEY 122

Strong Role Differentiation: Military Ethics and the Ethics of Military Research. *By* CAROL ANN SMITH 131

Ethics and Nuclear Weapons Research. *By* PAUL S. BROWN 136

The Ethics of Corporate and Academic Commitments to Military Research. *By* JOSEPH F. COATES 149

Ethics and Biological Warfare Research. *By* LEONARD A. COLE ... 154

Ethical Elements of Chemical/Biological Defense Research. *By* SHIRLEY A. LIEBMAN 164

The Place of Department of Defense-Sponsored Research at the University. *By* JOSEPH P. MARTINO 172

Ethics in the Pursuit of Ethics. *By* DANNY COHEN 184

Military Research, Secrecy, and Ethics. *By* VIVIAN WEIL 193

Scientific Freedom and Scientific Responsibility. *By* HAROLD C. RELYEA 200

Engineering Ethics and the Question of whether to Work on Military Projects. *By* STEPHEN H. UNGER 211

Militarism and the Quality of Life. *By* ALEX C. MICHALOS 216

Economic Conversion: An Alternative to Military Dependency in the University. *By* JONATHAN FELDMAN 231

Part IV. Reconsiderations of the Strategic Defense Initiative

The Ethical Imperative for Limited Strategic Defense. *By* HARRISON H. SCHMITT 245

Reflections on the Moral Debate over the Strategic Defense
Initiative. *By* STEVEN LEE 254
Strategic Defense Initiative Research: A Gray Area for Ethics.
By AANT ELZINGA 262

Index of Contributors 277

Financial assistance was received from:

- John D. and Catherine T. MacArthur Foundation

The New York Academy of Sciences believes it has a responsibility to provide an open forum for discussion of scientific questions. The positions taken by the participants in the reported conferences are their own and not necessarily those of the Academy. The Academy has no intent to influence legislation by providing such forums.

Preface

The institutionalization of modern science and technology can be divided conveniently into three stages. In the first stage, modern science was, during its rise in the seventeenth century, an activity of upper-class individuals. The Royal Society, founded in 1660, was primarily an association of such individuals.

In the second stage, during the nineteenth century, science became institutionalized in university departments and industrial research. The first degree in science, for instance, was awarded by the University of London in 1860. This gave rise to the professional organization of science.

During both of these periods, the ethical governance of science could rest with individuals and independent professional organizations. Traditional ethics commonly addresses itself to problems associated with personal morality and social-role responsibility in ways that make it possible to include scientific behavior within the common moral discourse.

The Second World War, however, brought science and technology in the United States unprecedented military funding, and introduced a new level—the third stage—of institutionalization. Aware of the distorting influences inherent in such largess, the scientific and technological community strongly supported the postwar creation of the National Science Foundation as an alternative source of governmental money.

Four decades after its founding, however, the National Science Foundation remains, in comparison with the Department of Defense, a small source of scientific research funds. The Department of Defense continues to control close to 70% of federal funds expended for research and development; the National Science Foundation, approximately 3%. But during the last forty years the scientific community has continued to raise ethical and policy issues about the military funding of a number of specific projects, from thermonuclear, biological, and chemical weapons to antiballistic missile and strategic defense systems.

In the belief that it was appropriate to consider the issue of the military funding of science from a larger perspective, while trying to compare and contrast particular positions developed in different specific debates, the Philosophy and Technology Studies Center of Brooklyn Polytechnic University and the New York Academy of Sciences organized a three-day conference entitled "Ethical Issues Associated with Scientific Research for the Military," of which the present volume is a slightly abbreviated proceedings.

The intention, it is crucial to note, was to steer away from the more common science–policy discussions in the interest of bringing together working scientists–engineers and philosopher–ethicists to grapple with issues of applied ethics relevant to the large-scale institutionalizations of science and their relation to military funding. The implicit thesis is that such engagements call for new reflection on the ethical dimensions of scientific and technological research.

INSTITUTIONAL BACKGROUND OF THE CONFERENCE

In May 1986, the New York Academy sponsored a conference on "The High Technologies and Reducing the Risk of War." The aim was to describe new defense-related technologies and to clarify their strategic and political implications as well as the problems and prospects for arms control.

During the preparations for this conference, in the fall of 1985, the Science and World Stability Study Group of the New York Academy began to explore the possibility of a further discussion of moral and ethical issues surrounding high-technology military research. At the same time, the Philosophy and Technology Studies Center of Polytechnic University independently began to formulate similar plans. When the two efforts were made aware of each other by New York Academy President Kurt Salzinger, Carl Mitcham, the director of the Philosophy and Technology Studies Center, was invited to attend a meeting of the Study Group.

After due consideration, a Study Group consensus encouraged Carl Mitcham and Study Group member Philip Siekevitz, a neurobiologist at Rockefeller University, to develop a formal proposal for a conference on "Ethical Issues of Military Research." (Note that the title was refined over time.) Such a conference would extend the New York Academy tradition of complementing its scientific and technical conferences with symposia on a variety of social and public issues, including ethics. Specific examples are the conferences published as *The Social Responsibility of Scientists* (*Annals* volume 196, article 4, 1972), *The Social Responsibility of Engineers* (volume 196, article 10, 1973), *Ethical and Scientific Issues Posed by Human Uses of Molecular Genetics* (volume 265, 1976), *The Role of Animals in Biomedical Research* (volume 406, 1983), and *Biomedical Ethics: An Anglo-American Dialogue* (volume 530, 1988).

A conference on ethical issues in military research is likewise a natural outgrowth of the New York Colloquium on Philosophy and Technology, sponsored since 1983 by the Philosophy and Technology Studies Center. New York Colloquium conferences have previously addressed "Information Technology and Computers in Theory and Practice" (1983) and "Phenomenology and Technology" (1986), with special lectures and symposia on such topics as women and technology, computers and Soviet society, biomedical ethics, and the ethics of the Strategic Defense Initiative (SDI).

This conference was designed, however, not just for professional philosophers or scientists—but for both. The understanding of "philosopher" is also intentionally quite broad, including scholars from a number of disciplines (particularly the social sciences) concerned with ethical issues.

PREFACE   xi

Neither was the purpose to take a partisan stand or to debate once again the pros and cons of specific weapons research projects, for example, SDI—although in the end there was a desire to review the recent SDI debate. Instead, the goal was to bring scientists and engineers representing diverse aspects of the political–moral spectrum in the SDI and related debates together with philosophers and others who can help place such debates within a larger context.

The conference ultimately took shape with the guidance of a subcommittee of the Conference Committee of the New York Academy of Sciences (Pierre Hohenberg, Belcore, Inc.; Melvyn Nathanson, Rockefeller University; Robert Weingard, Rutgers University; and Victor Wouk, Victor Wouk Associates) and an advisory committee for the Philosophy and Technology Studies Center (Paul Durbin, University of Delaware; Douglas Lackey, City University of New York; Steven Lee, Hobart and William Smith Colleges; and Vivian Weil, Illinois Institute of Technology).

PRECEDENT AND NEED

The issues outlined for this conference have not previously been addressed in any substantial or systematic manner. Indeed, they have scarcely been addressed at all. There has been no major conference, monograph, or article devoted precisely to this kind of moral reflection on the personal issues related to military-funded scientific and technological research. Previous discussion has focused either on some specific kind of military research (for example, nuclear weapons, biochemical weapons, and strategic defense), on the more restricted issues associated with scientific freedom, or on the character of the public policy process. Even in discussions of the morality of various public policy decisions, very little has been done to explicate and analyze the kinds of ethical arguments to which individuals and groups appeal. Even less has been done to examine the ethical problems of individuals and the questions they raise.

In developing this proposal, a further aspect of the need for such a conference became known. The program on Ethics and Values in Science and Technology at the National Science Foundation has for over a decade been a primary source of support for discussion of the ethical issues associated with science and technology. Yet this program and its successor, the program on Ethics and Values Studies, are precluded by their guidelines from funding programs "focusing solely or primarily on ethical issues associated with military technology and national defense strategies." This means that not only is a primary form of mission-oriented research placed out of bounds for ethical analysis, but also that a common justification for the National Science Foundation itself and especially some of its projects—that they contribute indirectly to national security by keeping the United States ahead of others in pure research—cannot be effectively broached. Without prejudice regarding ethical judgments concerning the association of science and technology with military needs and ends, there are clearly a number of important issues to be raised and addressed—issues that Ethics and Values Studies is not able, and other government agencies (for example, the Army Research Office, Office of Naval Research, Air Force Office of Scientific Research, Defense Advanced Research Projects Agency, National Aeronautics and Space Administration, and Department of Energy) would be unlikely, to consider.

Given this situation, we are thus especially grateful that the John D. and Catherine T. MacArthur Foundation agreed to provide the basic funding to make this conference possible.

THE CONFERENCE ITSELF

The conference opened with a review of New York Academy efforts in the area of ethics by scientist Philip Siekevitz, a provocative challenge by engineer George Bugliarello, and a conceptual overview of issues by philosopher Carl Mitcham. This led into the presentation of general ethical arguments for and against the military support of science. Robert Dinegar, an ordnance chemist from Los Alamos National Laboratory, offered the pro presentation; Bernard Roth, a mechanical engineer from Stanford University, argued against. Somewhat surprisingly, Dinegar gave not a utilitarian but a much stronger deontological–religious argument for the inherent rightness of the connection between science and the military. By contrast, Roth developed a rather subjectivist position on the basis of a personal dislike for the military. Commentaries by Douglas MacLean, Roger Shinn, and Paul Durbin sought to relate these and other possible pro and con arguments to the traditions of secular ethics, religious ethics, and the philosophy of science and technology, respectively.

The next two days included presentations on the political and historical background (by John Holmfeld, Harold Relyea, Alex Roland, and Rosemary Chalk), as well as presentations on conceptual issues concerning the definition of military research and the inherently ethical character of science (by Rustum Roy and Kristin Shrader-Frechette). Rachelle Hollander of the Ethics and Values Program at the National Science Foundation reviewed some related studies. Philosophers and ethicists Deborah Johnson, Douglas Lackey, and Vivian Weil developed ethical criticisms of the science–military connection, while Kenneth Kemp, J. Bryan Hehir, and Carol Ann Smith provided various rationales. Scientists Paul Brown, Joseph Martino, Joseph Coates, Danny Cohen, Erich Kunhart, and Stephen Unger variously defended work for or opposed to the military, and challenged their critics. Alex Michalos and Jonathan Feldman provided alternative rationales and funding strategies for scientific research. (Because they chose not to write up their presentations, the contributions by Holmfeld and Kunhart do not appear in this volume.)

The temper of the discussions provoked by the conference can be imagined by considering, for example, the presentations by Leonard Cole (social scientist, Rutgers University) and Shirley Liebman (biologist, U.S. Army Chemical Research, Development, and Engineering Center, Aberdeen Proving Ground), who provided strongly opposed views of biological weapons research. Persons from the military also attended and took an active role in discussions. Given such topics and circumstances, we expected a sharp exchange of views, and we were not disappointed.

By way of conclusion, there was a retrospective look at the moral debate over SDI by Harrison Schmitt, Steven Lee, and Aant Elzinga.

Among the presenters there was a rough balance not only between philosopher–ethicists and scientist–engineers, but also among those in various ways opposed to or supportive of the military support of science.

Three themes recurred throughout:

- The need for more careful conceptual distinctions when talking about the military funding of science.
- The responsibility of critics to make careful distinctions among various kinds of research when challenging scientists working for the military.
- The complex character of professional ethics in a military setting.

For example, would not even a strong antiwar scientist have to support research on bulletproof vests? Some critics further fail to utilize the most accurate information or make the most charitable assumptions about the positions of those they criticize. In many cases, it is also difficult to discern the practical applications of basic research. Given the fact there is much more "spin-in" that "spin-off" in the relation between military and civilian science, there could be a significant part of science that any consistent opponent of the military would have to avoid, no matter what its funding source. Although their suspicions are often unfocused, many scientists who oppose military-funded research clearly have some basis for distrust.

It was also clear that in most instances presenters could do no more than begin to initiate an inquiry, and that connections between topics deserve further examination. For instance, the question of the absence of any reflection on the role of the military in the standard philosophies of science (which itself was little more than hinted at by Durbin) is surely related to the ideological use of the idea of the value-free character of science and technology (a partial focus of the paper by Shrader-Frechette) in ways that went unremarked.

Precisely because of the richness and importance of the issues, there was a strong consensus that although such interdisciplinary and interethical exchanges are inherently difficult, they deserve to be promoted.

Two persons deserve special acknowledgment for contributions to the conference and the editing of this proceedings volume. Ms. Carla Manzi of the Conference Department at the New York Academy handled conference logistics with admirable aplomb. Mr. Kevin Mayer did an outstanding job on bringing rough manuscripts into publishable form and seeing this volume through the press. Indeed, Mayer's keen eye and deft editorial pencil helped many authors, including ourselves, to achieve a facility of expression that would have otherwise eluded us. Insofar as such facility has continued to elude us, of course, it is no fault but our own.

CARL MITCHAM
PHILIP SIEKEVITZ

Foreword

The Genetic and Psychological Basis of Warfare as a Challenge to Scientific Research

GEORGE BUGLIARELLO

*Polytechnic University
Brooklyn, New York 11201*

The use of scientific research for military purposes goes back a long time. Archimedes in Syracuse, and the anonymous inventor of poison gas in the China of the Sung dynasty, were early examples of scientists involved in military research. Ever since, the connection between science and military power has been obvious and strong. Suffice it to think of Leonardo's war machines or of Prince Henry the Navigator's and the British Admiralty's quest for better navigation techniques. Such techniques—as with Archimedes' levers or solar furnaces—also had a strong civilian application. In fact, it is often hard to decide which was the prior event—the military or the civilian. The difficulty of the separation is particularly striking with the development of explosives. Suffice it to think of Alfred Nobel's invention. But with the development of more sophisticated weapons in World War I—including airplanes, tanks, submarines, and poison gas—and above all with the development of rockets and nuclear weapons in World War II, there emerged research and scientists in large numbers focused exclusively on military purposes—even if, again, their inventions often had civilian applications.

Hiram Maxim's machine gun, with its rapid killing power, was perhaps the first modern weapon to arouse widespread ethical concerns. These concerns became enormously magnified with nuclear weapons. The prospect of wholesale destruction of hundreds of millions of human beings generated a horrified universal reaction that properly has not abated. On the other hand, nuclear weapons, far more than any other kind of weapon, became a successful instrument of geopolitical stability and discouraged large-scale war; the utility of nuclear weapons continues unabated, as long as they are in the hands of responsible powers. They have, in effect, provided a new twist, a technological one, to the old saying *si vis pacem para bellum*—if you want peace prepare for war. And in so doing, nuclear weapons have also provided us with difficult moral issues. To put it very bluntly, if we accept the fact that the human race seems to continue to be constitutionally prone to violence and aggression and tends to resolve many of its problems by resorting to force, is it unethical to do military research that helps maintain stability (even if not, necessarily, justice)? When might it be proper to object to a particular kind of research that is being carried out, and to how it is being carried out? I believe that this second question—that of limitations and directions of research—rather than the question of whether there should be any military-oriented scientific research, deserves at this moment the concentrated attention of the scientists, the military, and the community in general. Namely, if the saying *si vis pacem para bellum* is accepted as a fact of life, what kind of military-oriented scientific research can lead to preservation of peace or large-scale stability, and yet limit genetic damage and suffering? Also, what are the ethical ways in which scientists who disagree should do so individually or collectively, without weakening the essential goal of stability?

But it may be said that we should not accept in perpetuity those physio-psychological traits of the human race to which I have just referred—namely, the constitutional propensity to violence and aggression, and the tendency to resolve problems by resorting to force—traits that may have served us well when we were few and primitive, yet now threaten our very survival. Military research will continue to be necessary until we succeed in changing those traits. But for the first time in the history of war, thanks to advances in the cognitive sciences and in genetic research (for example, the prospect of complete human genome mapping), we have a serious if distant hope of finding ways to change some of our dangerous ancestral traits. Perhaps the most important ethical issue in connection with military research is how to press vigorously, as a correlate of military research, for research focused on reducing human aggression.

Research into the genetic and psychological basis of warfare is, I submit, a new and crucial obligation of science. There are strategic defenses not only of high-energy physics, but perhaps also of biology, and these biological defenses may be directed against our own most self-destructive instincts. Although such research and any possible applications must be undertaken with the strongest ethical controls and with a commitment to public education (the latter being necessary to ensure the integrity of democratic decision making), this research promises the possibility—albeit distant—of transforming the science–warfare relationship more fundamentally than anything in the past. In so doing it brings in the dimension of ethics in even more profound and necessary ways than nuclear weapons. It is a subject I commend to the attention of this timely and groundbreaking conference.

PART I. MAIN THEMES

The Spectrum of Ethical Issues Associated with the Military Support of Science and Technology

CARL MITCHAM

Polytechnic University
Brooklyn, New York 11201

Philosophical and ethical issues associated with the relations between science, technology, and the military have been largely ignored by the contemporary professional philosophical community. This is not so true with regard to philosophers in the past, particularly political philosophers, or with nonprofessional philosophers in the present. For example, physical and biological scientists who have been involved in military or military-funded projects, as well as social scientists (especially psychologists, economists, and historians), have initiated a number of relevant discussions.

The discipline of philosophy can, however, contribute to existing discussions by 1) offering conceptual clarifications of some empirical questions; 2) providing a framework for general interdisciplinary reflection; 3) promoting the synthesis of relevant information and arguments; and 4) offering criticism of some theoretical presuppositions entertained and positions advanced. To initiate such contributions, this introduction will indicate something of the general historicophilosophical background, then turn to a quick review of more immediate post-World War II discussions. Along the way it will indicate some philosophical, often ethical, questions that deserve to be addressed. A conclusion will consider the role of ethics in its various dimensions.

GENERAL BACKGROUND

Warfare is commonly divided into historical periods on the basis of its distinctive technologies. Modern military history, for instance, is ushered in by the use of gunpowder and the printing press (to disseminate nationalist propaganda); recent military history begins with the introduction of atomic weapons and jet aircraft at the end of World War II.

If the tables are turned, the history of science (commonly periodized by the prominence of distinctive theories) and the history of technology (easily periodized by its material and power sources) can be separated into different eras on the basis of differential engagements with warfare. In what can be called the first wave of modernity, modern science and modern technology (as distinguished from premodern science and traditional technology) are commonly imagined in terms of war and violence. Machiavelli (the father of modern social science) defends his positivism with the suggestion that fortune is a woman whom, if she cannot be seduced, can at least be raped. According to Francis Bacon (one of the founders of modern natural science), nature reveals her secrets more readily on the rack of experimentation than when left free and at peace, and is subject to scientific conquest. And according to Descartes, human domination is to be asserted over nature through the new way of (scientific) thinking that he wishes to introduce.

Such images remain alive in contemporary references to scientific "wars" on polio, small pox, cancer, hunger, or the acquired immune deficiency syndrome. Indeed, some scientists and science policy advisors have suggested that only fear of war can galvanize a democratic public to contribute to science with sufficient generosity to create a world-class scientific establishment. Certainly strong *prima facie* evidence exists that the most outstanding science is in fact associated with world-class military power (the United States, the Soviet Union, and Europe), and that attempts to fund science through nonmilitary projects result in smaller percentages of the gross domestic product being devoted to pure research or first-class science.

As has been argued by Charles Boyle, a physicist at Trent Polytechnic in Nottingham, "The military applications of science and technology form no mere fringe activity indulged in by a few atypical tinkerers and experimentalists. In financial and manpower terms, they are central to modern research and development."[1] Not just nuclear power but also drugs, pesticides, aircraft, radar, processed food, satellites, computers, transistors, lasers, and many other technologies have all been funded and developed primarily by the military for military purposes—although often with collateral justifications that emphasize their civilian spin-offs or applications.

Such arguments, however, indicate a strong need for conceptual clarifications and distinctions between research and development, between pure and applied or mission-oriented science, between science, engineering, and technology, and between military and civilian science-technology, in order to interpret and assess accurately a wealth of conflicting claims that can be marshaled from the available evidence in this area. The epistemology and philosophy of science might well address the issue of the adequacy and accuracy of the science-as-war metaphor, as well as possibilities for alternative forms of science. And the ethical legitimacy of using war fear or the war metaphor to promote and fund science, even for civilian peacetime benefit, is surely an issue deserving scrutiny.

In the second wave of modernity, the Enlightenment elaboration of the science-as-war metaphor was restated as the theory that warfare of human beings against each other could be replaced by a war of all human beings against nature. Hume and the *philosophes* advance such a view, arguing that human warfare is caused by material scarcity and the unenlightened pursuit of self-interest through a competition for resources. A better approach is for human beings to unite through industry, using science and technology to remove scarcity and thus the cause of internecine conflict. Appeals to the need to discover "the moral equivalent of war" (from William James to Jimmy Carter) are contemporary reformulations of such a view.

By contrast, Plato (in the *Republic*) and more recent historicophilosophical studies (see especially Lewis Mumford's *The Myth of the Machine*[2]) have advanced the thesis that warfare arose in association with the scientific and technological creation of "unnatural" or exceptional wealth and the resultant need to defend such wealth against those who "naturally" desire to share it. The first standing armies are associated with the fifth century B.C.E. transformation from hunting and gathering societies to domestic agriculture and the resultant surpluses. Whether such surpluses merely make possible a division of labor between agricultural workers and soldiers in a society already primed for war or actually provoke the rise of warfare and the military class is not at all clear. What is the relationship, for instance, between the ritualized fighting of archaic tribes and so-called civilized warfare? Such contrasts point toward fundamental questions about human nature on which philosophy might be able to cast some light.

In the third wave of modernity, which is associated with ideas developed by Nietzsche and elaborated upon by Freud, the idea that human Eros or the drive for self-preservation is complemented by Thanatos or a self-destructive death wish introduces another dimension to this discussion, one that was actually foreshadowed in the reactions of scientists and engineers to the destructive applications of their research and development. Rudolf Diesel, for instance, the inventor of the Diesel engine, committed suicide in 1913 in despair in part over the inability of technology to effect true human solidarity; his son, Eugene Diesel, under the influence of the experience of World War I, turned away from engineering and spent his life pointing out the social and cultural violence associated with technology.

The counterargument of Alfred Nobel, another inventor, and military theorist Jan Block, is that increasingly advanced weapons would make war so horrible it would become unthinkable, and that modern weapons would thus promote solidarity by a kind of technologically necessitated repression of the aggressive instinct. Although this theory has been to some extent confirmed by the post-World War II nuclear peace, it could also be argued that a "return of the repressed" has been manifested in proxy wars of so-called national liberation, in terrorism—and perhaps even in the worldwide assault on the environment that has created acid rain, oceanic pollution, rain forest destruction, and ozone depletion.

Indeed, today it is the disenchantment caused by the "side effects" or "unintended consequences" of the scientific and technological conquest of nature that is doing more to unite humanity than any positive vision of possibilities put forth by Enlightenment thinkers.

POST-WORLD WAR II DISCUSSIONS

Against such a general background, it is perhaps useful to consider as well the more immediate post-World War II science and technology policy debates in the United States.

Before World War II, despite the existence of some national arsenals and military research facilities, science and technology were primarily funded either by private foundations (mostly of a medical sort), by private enterprise (for example, the work of Thomas Edison), or by civilian agencies of the government (for example, what are now the Department of Agriculture, the Department of Commerce, and the National Institutes of Health). As a result of the World War II "mobilization" of science and technology, however, the War Department—now the Department of Defense—became *the* primary source of funding for scientific and technological research.

To redress the balance, immediately after the war scientists themselves lobbied for the establishment of a federally funded, civilian-controlled foundation through which scientists, independent of the distortions introduced by military interests, might determine scientific funding priorities. The founding legislation for the National Science Foundation (NSF) nevertheless proclaims in its preface that the NSF exists to promote not only "scientific progress" but also "national defense," and it is debatable to what extent it has developed a true independence of military concerns. Again, there is a need for conceptual analysis and clarification of the relations between different types of science and technology, at epistemological, metaphysical, and sociological levels.

At the very least, NSF funding of science–technology has failed to approach anything like the amount of funding provided by the various agencies of the Department of Defense (for example, the Office of Naval Research and the Army Research Projects Agency). Between 1950 and 1985, 65–70% of federal research and development funds were channeled through the Department of Defense, only 1–3% through the NSF. (If one includes the Department of Energy, whose major focus is nuclear weapons, and the National Aeronautics and Space Administration, which is under heavy contract to the military, the military-related totals go even higher.)

In this relationship between science–technology policy and the military, some comparisons between countries can also be informative. Although the United States invests about 5% of its gross domestic product (GDP) in the military, while the Soviet Union invests over 10%, the absolute figures are about the same, because the GDP of the United States is roughly twice the GDP of the Soviet Union. Although such statistics are subject to considerable debate because of the ways different countries use different reporting categories, figures provided by the Stockholm International Peace Research Institute can be interpreted to indicate that Britain and France, like the United States, devote roughly 5% of their GDPs to the military or military-related expenses.[3] Indeed, the world totals for all countries may be at about this same level.

Yet given the high-tech bias of U.S. military policy, a greater proportion of the U.S. military expenditures inevitably go to science–technology. The policy rationale is, as is well known, that a democracy simply is not capable of making the kinds of human sacrifices necessary to oppose in kind the large conventional military establishment of the Soviet Union, which must therefore be countered with more sophisticated weapons.

One result of this policy, it has been argued, is what is known as the arms race and the development of ever newer generations of ever more sophisticated weapons: generation after generation of nuclear weapons, missiles, antipersonnel bombs, chemical and biological weapons, detection and communication electronics, spy and interference electronics, lasers, precision guided munitions, and now smart or robot weapons. The moral legitimacy of a policy that promotes and rationalizes such development, even in the name of democracy, should be open to philosophical examination. The readiness with which science and technology are captured by such a policy may raise epistemological and metaphysical questions appropriate to the philosophies of science and of technology.

In response to the implications of such questions, it might be pointed out that the creation of ever more high-tech weapons actually began with the industrial age. The U.S. Civil War, 1861–1865, introduced torpedoes, mines, grenades, the rifle, machine guns, balloons, iron ships, and submarines—in what was a more intense period of military invention than ever before and maybe since. Refinements continued to be made, of course, and in World War I tanks and aircraft were introduced. But it was not until World War II, and the advent of the nuclear age, that weapons of an entirely different order again appeared (although Germany developed the V-1 and V-2 rockets, precursors of today's cruise and ballistic missiles, it was the United States that developed the first atomic bomb). How significant is it that the United States led in applying truly novel technology to weapons production at both the advent of the industrial age and the advent of the nuclear age? Is it possible that the U.S. democratic culture is inherently more prone to this kind of weapons development?

At the same time, the United States has witnessed a number of debates among scientists about particular weapons systems, from the hydrogen bomb to chemical and biological weapons and the Strategic Defense Initiative. These different discussions deserve to be compared and contrasted in search of common themes and issues. For example, the restrictions of defense contracting and the need for secrecy in many aspects of weapons research have often been criticized as being opposed to the ethics of science and the ideal of the free exchange of information. In one instance, Norbert Wiener, the inventor of cybernetics and a pioneer of information control theory utilized by radar, came to a conclusion not unlike that of Archimedes, that he would no longer publish any scientific research that could be put to use by the military, because it would inevitably be misused. For Wiener, the ethics of science required the withholding of information, not its free exchange. Has such an argument occurred within debates related to chemical or biological weapons? Are there common responses?

AN OVERVIEW OF PHILOSOPHICALLY RELATED ISSUES

In light of the general background and more recent science policy discussions, it is possible to indicate in summary fashion a spectrum of philosophically related issues to be addressed by interdisciplinary reflection among scientists and engineers, together with philosophers, historians, social scientists, public policy analysts, and others. To be considered would be 1) conceptual distinctions between pure and mission-oriented research and possibly among different specific missions; 2) general analyses of the influence of the military on science–technology and the sociohistorical background of past and current debates on military-supported research projects; 3) the kinds of arguments employed in the articulations of different positions; and 4) the moral options open to persons on various sides of the issues.

In the course of exploring these four realms of reflection, the following kinds of more specific questions would be addressed: Conceptually, what differences exist between mission-oriented (applied) and nonmission-oriented (pure) research? Between military and nonmilitary missions in research? Between direct and indirect consequences and problems? Does the scientist or engineer doing mission-oriented or military research have different political and moral responsibilities than one doing nonmission-oriented or nonmilitary research? Are there distinctions to be made among various kinds of scientific and technological research? What are the corporate and academic relations to military research?

In provisional clarification, it is important to note, for instance, that the military support of science and technology is not limited to funding the research and development simply of weapons or weapons systems, whether defensive or offensive. One can imagine a variety of nuances in pro and con attitudes toward the military funding of a) research related to offensive weapons, b) research related to defensive weapons, and c) research not directly related to offensive or defensive weapons.

Historically and sociologically, what have been the real-world imperatives associated with scientific research? To what extent have military needs always exercised an influence over the funding of scientific research and scientific progress? Are there differences between the military research conducted at universities, at independently established government laboratories, and at private-sector research institutions? How have scientists and engineers responded in specific situations? What kinds of general arguments have they made? How is the military engaged with science–technology policy in different countries? What have various moral philosophers (and scientists) had to say about ethical issues in relation to military research? What has been the social context of different discussions of ethical issues involved with scientific research? In what various ways have support for and protest against military research been practiced? How might psychological development be related to moral analyses of the issues at hand?

Agonisticly, is warfare justified or necessary? Might it be an ontological component of being-in-the-world? What are the strengths and weaknesses of various ethical theories (for example, utilitarian, deontological, natural law, and virtue ethics theories) in dealing with the moral issues of military research? Do some ethical frameworks lend themselves to justifying and others to criticizing military research? What is the relation of just war theory to military research? Are there ethical issues in this area that are not adequately addressed by traditional ethical theories? What is the proper role of secrecy in science? In technology? What is the relationship between the moral arguments associated with public policy debates and those having to do with personal ethical decisions?

In relation to practical action, do the professional ethics of either scientists or engineers have anything special to say about military research and development? Does the social status of the scientist or engineer alter moral responsibilities or obligations? What are the forms of action appropriate to different kinds of moral judgments of military research? How can individuals develop and promote their ethical understandings? Does ethical discussion really make any difference? Do military researchers acquire special legal responsibilities? What kinds of groups exist for the support of ethical responsibility among scientists doing research sponsored by the military? Does the nature of the political system (democratic or totalitarian) determining research priorities affect the moral responsibilities of the individual researcher?

THE ROLE OF ETHICS

In conclusion, the relation between science, technology, and the military can be observed to engage three different but not always distinguished senses of ethics. The sketching of such distinctions also allows for some preliminary assessment of the relative weights of different ethical issues and the differential engagements of various philosophical arguments.

Insofar as ethics is concerned with the problem of *acting*, or doing what one thinks is right in the face of countervailing pressures, especially those of an economic character, the military funding of science and technology obviously introduces weighty problems. Numerous scientists, especially those working at national laboratories such as Lawrence Livermore or Los Alamos have experienced serious crises of conscience and have been forced to grapple with the making of some basic changes in their research and institutional commitments. (The case of Peter Hagelstein from Lawrence Livermore is a recent case in point.[4]) The need for exercising moral courage is often manifest. This is also true for those scientists who believe in the importance of doing research at the university, where there is strong antimilitary sentiment.

At the same time, because the scientist and engineer is so respected and so employable in a wide range of fields in contemporary North America, it is doubtful that someone working at a national laboratory, a person already recognized as having superior ability and education, would have difficulties finding alternative employment. (Hagelstein, for instance, left Lawrence Livermore, but continued his career at the Massachusetts Institute of Technology.) Equal opportunities exist for scientists to move from civilian universities to the over one hundred research centers directly supported by the Department of Defense. Thus, although this area of ethics is important, it does not necessarily exhibit overwhelming urgency. For example, scientists doing military research tend to respect dissenters, and do not have to answer to any strong professional certifying body comparable to the associations found in medicine, law, and engineering.

Insofar as ethics is concerned with *deciding* what is right or good, there is both more confusion and more urgency. The need to decide anew, in scientifically and technologically changed circumstances, about how to apply traditional moral principles, is the most dynamic aspect of the contemporary movement known as applied philosophy—which includes, for example, biomedical ethics, environmental ethics, engineering ethics, and computer ethics. The ethics of the military funding of science promises to become a significant new addition to such interdisciplinary philosophical reflections.

Virtually all the moral issues indicated in passing above come into play here and involve conflicting implications and analyses, with cogent arguments on more than one side of a question, and perhaps even necessary trade-offs in values. Does military research really compromise scientific principles or values? Does it advance science, the military, or some complex combination of both? What would take place independent of military funding anyway? How much does the democratic process alter personal scientific responsibility? How much should scientists and engineers themselves contribute to the democratic and policy-making process?

Even the conceptual issues related to distinctions between science, applied science, technology, research, and development have applied ethical implications. Proximate and remote commitments and implications influence degrees of responsibility and ethical decision making.

Insofar as ethics is concerned with *knowing why* or with the grounding of decisions in general principles, discussion necessarily becomes more abstract. This nevertheless remains an important, even fundamental area of reflection, in which one can note certain differential engagements.

For instance, although it is certainly possible to imagine a deontological (or even ontological) argument for the military support of science and technology, it is difficult to find such an argument seriously presented in the available literature. Very seldom does one justify the military funding of science by arguing that warfare is an inherent feature of action or an essential feature of reality (but see Heraclitus, fragments 53 and 80). The most common justification is, instead, probably a utilitarian one. The military support of science is commonly defended not so much as a necessary duty but as an unfortunate necessity, and a choice of lesser over greater evils, or because of the benefits that it brings. The increased prominence of mission-oriented research in science, especially military-funded science, also serves to favor utilitarian over alternative ethical theories.

By contrast, one can readily conceive of a deontological argument against the military funding of science, or at least a deontological argument against scientists accepting military funding. Indeed, it would appear that those who most strongly oppose the military funding of science, as with principled pacifists, do so precisely on deontological grounds. One could also readily imagine a natural law argument against military funding, especially in that new version of natural law known as ecological ethics.

But, once again, conceptual distinctions must be brought into play, because there are differences to be considered between military and scientific-technological responsibilities in relation to such deontological and natural law claims. The differential implications of virtue ethics, rights theory, and feminist moral criticism must also be considered. But the investigation of such engagements could well throw light not only on the science–technology–military relationship but also on some of the fundamental principles of ethics, at least as these are relevant to the high-tech and militaristic world in which we now live.

REFERENCES

1. BOYLE, C. 1984. Military technology. *In* People, Science, and Technology: A Guide to Advanced Industrial Society. C. Boyle, P. Wheale & B. Stulgess, Eds.: 151–168. Barnes & Noble. Totowa, NJ.
2. MUMFORD, L. 1967–1970. The Myth of the Machine. Volume 1: Technics and Human Development (1967). Volume 2: The Pentagon of Power (1970). Harcourt Brace Jovanovich. New York, NY.
3. Stockholm International Peace Research Institute. 1980–present. Yearbooks.
4. New York Times. 1986. A laser's inventor ending arms work. September 11: p. A18.

The Moral Arguments for Military Research

ROBERT H. DINEGAR

Science Division–Chemistry
University of New Mexico at Los Alamos
Los Alamos, New Mexico 87544

INTRODUCTION

Discussing the moral arguments for military research is a tremendous task. Not only is the subject matter complicated, but any conclusions are of the greatest importance.

Specific answers will be difficult and perhaps even impossible to find. But this need not prevent us from thinking about the questions involved, nor should it discourage us from hoping that eventually the correct answers will be given to us. Meetings of the kind preserved in this volume may offer no end of help.

I shall attempt to discuss the morality of military research in the following manner: 1) by sketching the evolution of physical-scientific research (for this kind of research is what generally constitutes military research) in ethical-religious terms; 2) by saying a few words about research in general; 3) by discussing the morality of selected fields of peaceful endeavor that issue from military research; and 4) by considering the problems of warfare in general and in the nuclear age, these problems being the focus of most military research.

EVOLUTION OF PHYSICAL-SCIENTIFIC RESEARCH IN ETHICAL-RELIGIOUS PERSPECTIVE

According to one scientific theory, many billions of years ago there appeared within the boundaries of the universe an extremely dense mass of elementary particles. Governed by complicated physical and chemical laws, this enormous quantity of matter proceeded to react and generate the building blocks that make up the elements of the world around us. In a consistent and repeatable way, albeit a complicated and not completely understood one, this original cloud changed into the mixture of fundamental particles necessary to form the atoms and molecules that constitute our physical universe.

Eons passed—probably seventy-five percent of the total time from the beginning until now—and living organisms appeared. Although composed of the same atoms and molecules as nonliving matter, these complexes were put together in a special way and endowed with special properties. The gift of life was given to the universe. Many thousands of years ago, the act that was begun in antediluvian ages reached fruition. Man appeared on the scene of history.

Here it is permissible to go beyond physical–scientific research as, contrary to the customary secular analysis, I propose to do. Fashioned in the image of the Creator of all things, man was assigned special duties. He was to have dominion over all that had been given. He was to be the Creator's vicar on earth, the Creator's steward. For this the Creator made order out of chaos.

The account of how man first accepted the honor and discharged the duty he inherited is told in story form in the third chapter of Genesis. This myth relates that instead of accepting and nurturing the great gift of life man chose to think of it as being no gift at all. Instead of humbly accepting his role as the one who receives and responds, he chose to think of himself as the giver. Man chose to deny his correct place in the scheme of creation. He used his God-given power of free will to alienate himself from his Creator. For this, he was banished from an earthly paradise.

But man was not destined to be cut off from his heritage forever. To aid him in his attempts to return to the better life, God continued to give him great gifts.

Over the same millennia in which man's supernatural and spiritual advancement took place, so did his natural and material advancement. Although not synchronous in development, no one aspect of man's being far outstripped the others. Just as the revelation of the one God was given by means of a slow development from local, tribal, and national deities, the crude Paleolithic and Neolithic ages did not last forever but were replaced by the Bronze, which in turn gave way to the Iron. Man passed from disorganized thought to the beginnings of a rational systematic attack upon the problems that surrounded him.

In the several centuries before the birth of Christ, the Greek system of logic was applied to the knowledge and ideas that had been accumulated in the more distant past. It is in these ages that we find the first discussions of the concept of the atomic structure of matter by Leucippus, Lucretius, and Democritus. Although these thoughts lay dormant as seeds for twenty or twenty-five centuries, many other discoveries of the first magnitude did germinate during these intervening years.

By the early part of the nineteenth century, men of great ability were turning their thoughts toward trying to understand the constitution of matter on more than a qualitative, philosophical scale. Credit for originating the first atomic theory of matter is usually given to John Dalton.

By the beginning of the twentieth century, the atom was known and accepted but little was known about it. The clear concept of the constitution of the atom was given in 1911 by Ernest Rutherford. During the first years of this century, Rutherford and the Curies (Marie and Pierre) changed the course of physics and chemistry by demonstrating that the unusual property of uranium discovered by Henri Becquerel in 1896—that of emitting radiation—was not only present in other elements but was also intimately connected with the fundamental structure of the atom. In 1913, Neils Bohr, then a young man working in Rutherford's laboratory, brought forth the first satisfactory theoretical model of the atom. In 1932, John Chadwick discovered the neutron, a particle with relatively heavy mass but zero charge. Knowledge of this particle made possible the investigations of Enrico Fermi and others on nuclear reactions.

Particular mention must be made of the period 1938-1942. During these years, the practical utilization of the huge amounts of energy released in some nuclear reactions, predicted by Einstein three decades before, was shown to be a distinct possibility. The work of Hahn, Strassman, Frisch, and Meitner on the artificially radioactive substances produced when neutrons impinged on a uranium nucleus, as well as the work of Fermi, Szilard, and Joliot on the feasibility of a neutron chain reaction, culminated on December 2, 1942 (beneath the west stands of the University of Chicago's Stagg Field) in the first controlled nuclear chain reaction. The secret of the energy of the universe was at last revealed. Man's knowledge and comprehension had moved onward and upward exponentially.

I have presented kaleidoscopic highlights in the development of man's spiritual and material state[1] for two reasons: first, to indicate the true place of man in the "cosmic drama" of life as that of actor or participant rather than as author or producer; and second, to emphasize that the spiritual and the material aspects of men's lives are inextricably joined together. Moral evaluation of any of mankind's activities should only be made with this background in mind.

MORAL STATUS OF PHYSICAL-SCIENTIFIC RESEARCH

If one views the field of physical-scientific research from the previously described vantage point, it is quite obvious that it must be considered a gift from God rather than a creation of man.

Because physical-scientific research is a gift from God, it is inherently good; the moral effects are not a property of the work itself. Just as man's actions are simply manifestations of the morality or immorality of the soul that is enfleshed, the effects of this research show only the morality or immorality of those who indulge in it.

The problem of the moral effects of physical-scientific research lies then in what man has done with it in the past, what he is doing with it in the present, and what he will do with it in the future. In a way, research can be likened to the standard of the ancient Byzantine Empire, the double-headed eagle. One head looks forward expectantly to the glories of right usage; the other head is turned aside so as not to gaze upon the tragedies that will accompany misuse. In the words of Arnold Toynbee: "The marvelous gifts of science impose a moral responsibility on their recipients. We must take care to use the gifts for life and good. If we fail in this, science is going to deal out death and evil!"[2] St. Paul wrote that we are "stewards of the mysteries of God" and reminds us that it is required of stewards that they be found trustworthy. In how man responds, that is, in how wisely he integrates the spiritual and material aspects of his life, lies the answer to the moral and ethical problems of the scientific age.

PEACE-ORIENTED PHYSICAL-SCIENTIFIC RESEARCH

To consider in detail the moral effects of peace-oriented uses of physical-scientific research, some sort of criteria must be set up against which the uses of the research can be judged. For me that standard is the Judeo-Christian ethical system. Within this system, to be good, right, and proper—that is, moral—a particular aspect of physical-scientific research must contribute to the advancement of man's whole being, his soul as well as his body.

For convenience, the many aspects of this research that are now engaging man's interest will be discussed under two main headings: fundamental research and technology. Both of these categories contain not only areas of research that result in an overwhelming preponderance of good but also those areas that must have a concomitant sense of Judeo-Christian stewardship to realize their full potential.

Fundamental Research

An obvious example of a field of research that brings forth an overwhelming preponderance of good is that of medicine. The good accomplished by the advent of preventative, diagnostic, and curative techniques made possible only by the results of physical-scientific research is so great that there can be no question of the morality of these techniques. Although the moral virtue of courage is certainly aided by these means, the real morality of the advances that fall in this category stems from the fact that these acts of healing manifest the cardinal virtue of love.

An example of fundamental research that needs the proper sense of stewardship to realize its full potential is the investigation of the structure and nature of matter and energy.

By nature man is an inquisitive creature. From his very beginning he sought to understand the world around him. The acquisition of new tools and new methods has only increased his efforts to view phenomena through glasses that were not tinted with superstition or purely speculative fantasy. For at least three hundred years he has been on the track of producing a systematic and correlated body of empirical knowledge upon which he could superimpose his intricate theoretical efforts. The methods of physical-scientific research have added substantially to this store of wisdom and understanding. These additional gifts of the Spirit have proven most fortunate. Certainly an increase in the virtue of faith can be expected to come from work that over and over again clearly manifests orderliness and reason—attributes of a Creator God. I agree with those who see in the expressions of the universal truths uncovered by all forms of fundamental research the type of intellectual achievements that mirror, albeit darkly, the rationality of the One who created man in His own image.[3] This type of thinking augurs well by indicating that the advancement of mankind in the future will be done with prudence.

Yet there is another side to this particular aspect of the physical-scientific research picture. This is the situation that results from our not seeing that the discoveries of physical-scientific research reflect the boundlessness of our created order and man's small but important part in its destiny. Suppose, through the sin of pride or other defect, man cannot comprehend the existence of supernature and the transcendent significance of it all. Is this of any moral significance? To me the answer is unequivocally that it is. I am sure, however, there are those who would disagree. Several times in the past years, the Reverend William G. Pollard, priest and physical scientist at Oak Ridge, Tennessee, has dealt with this problem in a way with which I think we should all be familiar. Built around the idea that "the rise of science has brought with it an ever-increasing imprisonment of the mind and spirit of man in space, time, and matter," the thesis is put forth that our great interest and involvement in newfound sciences are excluding all other aspects of reality from our being.[4] By considering only those aspects of reality that are capable of being tested by the rules of the physical-scientific world, man shuts himself off from the grace of God.

Technology

Let me now turn to the area of technology, that is, the area of applied research. Almost immediately one is struck by the enormous complexity of the web that technology is spinning around us. It would be impossible to consider all the aspects, so I limit the discussion to nuclear power.

The first and most obvious use of power is the same in the nuclear age as in any other—as a means of securing the necessities of life. Just as early sources of power (such as the strength of the human back and arm, captive animals, and later that power derived from steam and electricity) have always been applied primarily toward producing necessities (such as food and clothing), nuclear power has much the same direction of emphasis. In fact, it is not hard to believe that advances in upcoming years could lead to the greater production of goods so necessary in a world in which our ability to reproduce ourselves threatens to outstrip our ability to provide for the maintenance of life.

Yet only a small number of the societies today are fortunate enough to be able to avail themselves of this great potential. It is in the interrelationship of the fortunate with the less fortunate that the moral aspect of the use of nuclear power arises, for certainly the products themselves are beneficial in their entirety. The morality hinges upon whether the more fortunate will work as hard and as industriously to take these advantages to the less fortunate as they did for themselves. It will do us no good to concentrate knowledge of industrial science and technology in the mind of man if we dilute the effectiveness of its use by refusing to sacrifice traditional privileges and prejudices.

But even if we share our methods and our knowledge, the problem does not stop here. We must also give of our inspiration and enthusiasm so that the less fortunate will eventually be lifted to the place where they can be self-sufficient, not just the recipients of charity. Let us not inadvertently be guilty of contempt. In the nuclear age it is most important that we inculcate a dedication of the will at both the individual and collective levels.

A second direction of emphasis for nuclear power is toward the exploration of outer space. Although well publicized and quite glamorous, this also is not really a new direction of human effort. Man has long wondered about the heavens, dreamed about exploring them, and desired to conquer them. Only with the advent of nuclear power, however, has he had any real hope of ascending the stairway to the stars. Once again, to find the moral implications, we look not at the endeavor itself, for surely the desire to explore raises few moral problems as long as pride does not motivate the action. The real moral problems are related to how we react to what we experience and discover. For example, how will man respond to experiencing the immensity of space? Already we have had different reactions. One of our astronauts found it conducive to the strengthening of belief in God while another of a foreign nation used it to strengthen the position of disbelief. Years ago, C. S. Lewis clarified this apparent paradox by pointing out that "space travel really has nothing to do with the matter. To some, God is conceivable everywhere; to others, nowhere. Those who do not find him on earth are unlikely to find him in space."[5]

Finally, will we share any success we have with others, or will we find it expedient and seemingly necessary for our own protection to be exclusively possessive and greedy? Will we try to forget our differences with those who have challenged us in the adventure? Will we count ourselves privileged to be the stewards of new worlds and desire to share them with others? Will a shrinking universe be crowned with a better appreciation of the responsibility of man to man than has been evident on a shrinking planet? We must be on our guard, for the peaceful uses of man's abilities can lead to conflict almost as quickly as the nonpeaceful uses. "The future depends on people," Edward Teller said years ago. We should share his hope "that in the long run they are reasonable."[6]

CONFLICT-ORIENTED PHYSICAL-SCIENTIFIC RESEARCH

I have saved the most vexing of the moral problems of the scientific age—those associated with weapons and war—for the last. To a person who is unfamiliar with the history and background of the development of science, this might seem like an improbable or at least unimportant problem. "Surely no one would expend a significant amount of effort toward producing items of war," he might say. "It is not worth considering." Yet we all know only too well that a great deal of scientific effort has been devoted to the development of weapons, and we all know that the nuclear age dawned on mankind when the world was in a state of armed conflict and we thought it necessary to be at least copossessors of the ultimate dispensers of death and destruction. This is not to say that the nuclear age would never have come about without war, but its fruition was motivated by a nonpeaceful impetus.

With the world in turmoil and at times in open conflict, what attitude should we take today toward war and the knowledge and technology we have acquired over the past half century? Should we continue along past lines, praying that war will never come, but justifying our possession of the ultimate weapons by invoking the right of self-defense? Should we summarily eliminate all weapons stockpiles and stop all weapons production, development, and research, again praying that war will never come, while admitting that we have sinned in the past and resolving never to be guilty again? Or do our morals oblige us to find some other answer?

One must start by saying that the moral framework within which to view warfare today is, in essence, the same as that to be used with any type of conflict in any age. The same, that is, with one important modification: acts of war in our times demand a newer and stricter evaluation of all the consequences. For example, one could justify participation in a conventional conflict according to certain rules; however, one would not necessarily be able to justify participation in a nuclear conflict—even if the same rules were to be applied. This is so because the components to be taken into account (the accounting to be performed according to the rules) would have to be given different weights in the conventional and nuclear cases.

Let me suggest an analogy: given it is permissible to shoot a terrorist standing in front of a window in an occupied building, do the same set of rules allow the bombing of the building—with the probable injury of a large number of other people—just to get at him? The answer is most certainly no, not because the rightness or the wrongness of killing are different in the latter case, but because the weights given certain components (the amount and type of killing) are different.

JUST WAR

The history of man's search for the conditions under which he could fight morally is known as the development of the doctrine of just war. Its underlying bases can be attributed either to casuistry—the science of resolving questions of right or wrong—or to a simple application of the principles of Christian ethics. A case can be made for either viewpoint. The doctrine is generally considered to have six conditions in the form we are familiar with today: the first five of these have to do with going to war; the sixth places bounds on the methods that may be employed. These restrictions may be expressed as follows:

1. The war must be resorted to only after all other means of solving the problem have been tried and have clearly failed.
2. The decision to go to war must be made by a duly constituted authority.
3. The war must be to right a very great wrong.
4. The war must be to establish only that which is good.
5. There must be a real prospect of victory.
6. The conflict must be waged within the limits of love, prudence, and justice.

Since Augustine first sketched such a just war himself, many people have analyzed war according to these terms. Whatever our personal opinions of the just war formulation may be, we should take a look at what this line of reasoning proposes with regard to armed conflict in our day. Even if we do not accept the answer that an application of these principles generates, we are obliged to take a look at what they say to us.

Let me first consider those conditions that have to do with going to war. Obviously aggressive war is prohibited.

Nevertheless, from the earliest of times—in fact and fiction, as part of profane and sacred history, in written and verbal form—all of existence, supernatural as well as natural, has been viewed as a battlefield where opposing forces meet to decide the fate of countless beings. The ancient Babylonians not only had their wars with the Assyrians but also wrote about the conflict of good and evil between the invisible Marduk and Tiamat; the Persians, who subdued the hosts of the evil descendant of Nebuchadnezzar, Belshazzar, also believed in the inevitability of the defeat of their god of evil, Ahriman; and the Greeks, who conquered the Persians, told tales of cosmic genocide between Kronos and Uranus. In sacred literature, it is God in a warrior's armor who tells his name to Moses and who takes Jacob's sons and daughters as his people to deliver them by an overwhelming show of force against the Egyptians. The army of Israel is called the host of Yahweh and is one of the appointed means of action for the Lord's elect. Throughout the Old Testament, especially during the period of the Tribes, the Judges, and the Kings, one is given the impression that the force of arms is somehow right, perhaps even holy.

With a background that stretches so far into the past, it is not strange that this idea of forcible action is still attractive today. It is, however, a mistaken attraction, and there are many errors in this way of thinking. It should be discarded.

But what about a defensive war—an application of the basic principle that a person has a moral right to defend himself against unilateral, unwarranted, and unprovoked attack that is specifically intended to destroy him? I believe that such a conflict can meet the first five conditions for a just war. These five conditions, however, apply to going to war, not to the methods of war.

It is with the sixth condition for a just war, the condition for limiting conflict, that the specter of the nuclear war really raises its moral/immoral head. When one considers the destructive power of nuclear weapons, and one is trying to apply the criteria for a just war, one begins to feel that the nuclear age has made it impossible to fulfill at least two of the necessary conditions for moral fighting: namely, 1) that the cause must be proportionate to the evils brought about by the war, and 2) that the war must be conducted within limits.

In fact, one could go even further and argue that the very possession of nuclear weapons as deterrents should not be allowed because weapons are ineffective deterrents unless one can assume they will eventually be used—and we have already refused to accept the use of nuclear weapons. It would thus seem that participation in a large portion of today's military research is prohibited.

Clearly this type of analysis shows an attitude of morality or love. But does it deal with love in a complete sense? I do not think so. I think this conclusion is true only as far as it goes, and it does not go far enough. Love has the property that you must desire what is best for the object of your love. To allow the object of your love to manifest the basest characteristics of fallen human nature is not what is best for him, for you are at least encouraging him to destroy his own soul. Far from showing the Christian virtue of love, this type of action is closely akin to the sin of passive scandal. Far too often this oversimplification of love with the emphasis on the *selfish* rather than the *selfless* component leads to morally improper action. In addition, whether it is compatible with the need for prudence in our never-ending search for a way to change our "swords into plowshares and [our] spears into pruning hooks" is open to question. I agree with what the Israeli general said after a United Nations cease-fire resolution was put into effect: "Now is the time to put our swords back into their scabbards, not to beat them into plowshares and pruning hooks." I also agree with Dr. Norris Bradbury, second director of the Los Alamos Scientific Laboratory, who said that weapons laboratories will be around as long as the unhappy situation continues in which the United States needs to be able to bargain from a position of strength and not weakness.

The greatest weakness of the argument against possessing a nuclear deterrent, however, is that it advocates a type of love that blinds—blinds one to the problems of right and wrong that exist in the world. Christian action allows, even demands, the elimination of injustice, wrong, and all forms of evil. The late Paul Ramsey once asked whether the love that required the Good Samaritan to help the waylaid stranger on the road to Jerusalem also demanded that he not stand idly by if he happened along the road while the stranger was being attacked.

On the surface, a unilateral decision not to participate in any form of armed conflict might seem right. So it is, provided this will not inevitably lead to some form of evil triumphing. No action can be morally right that results in a moral evil. No nation can lay down its weapons in isolation. A nation taking such unilateral action might secure physical safety, but only at the risk of losing its soul; for if it should come under the totalitarian yoke, its citizens would soon lose all that makes a full life possible. Anthony Eden also spoke to this problem just before Christmas 1964 when he said that to let evil go unchecked would result in anarchy, and this could be as bad as nuclear war. If morality does not actively move against the forces of evil what is its value?

It seems to me the whole point revolves around what we are trying to do. If we are trying to check the spread of evil, we are in the right. If the only way we can do it is to have and possibly use nuclear weapons, our duty—painful, distasteful, and loathsome as it may be—is to do it. Our main moral objective is to stand for what is right.

The question that now arises is whether we are right. If the purpose of a nation is to serve others so as to extend freedom, must we not say that self-defense on the part of that nation is right? The road we must travel is no easy one. There are signs along the road to warn us to be on guard, for danger always lurks near. We will have to be prepared to fight for our right to proceed on our journey. Our common sense tells us that we will use the means of defense temperately, prudently, and mercifully, but courageously, with wisdom and fortitude.

But what if, in spite of our best efforts, the horrors of nuclear war come upon us? How are we to view the fact that life as we know it on earth could be easily snuffed out by thermonuclear weapons? From the Christian perspective, it may even be that God will allow the world to end this way. Indeed, if Christianity can only equate the preservation of the human race with the will of God, there is little difference between its teachings and secular pleas.[7] The Christian proclamation is not that mankind, individually or collectively, will go on in this life forever, but that a new relationship has been established between the Creator and creation, and that this new relationship is sufficient for man in whatever mode of existence he finds himself. There is a sense in which this may be true in other religious traditions and even in some secular ethical systems.

CONCLUSION

In their time of doubt about the future, the ancient people of Edom looked to their watchman, the man who looked out over the surrounding lands from his tower, to tell them what the approaching day held. The prophet Isaiah tells us that he said, in response to their queries: "I see the daylight approaching – a great day; but after that more trouble – night again." These words are prophetic for us today, too. Great good has come and still is coming upon us. But it also brings great danger and possible death. But the teaching of Christianity and of morality is that we should not fear death as the greatest danger.

REFERENCES

1. My presentation is indebted to that of another priest-scientist, W. G. Pollard. See POLLARD, W. G. The Cosmic Drama (faculty paper). National Council of the Episcopal Church. New York, NY.
2. See, for example, SCHRADER, R. Science and Policy. Macmillan. New York, NY; see also SABINE, P. Atoms, Men, and God. Philosophical Library. New York, NY.
3. Our Sunday Visitor. 1963. **52**(27).
4. POLLARD, W. G. Christianity in the Space Age. *In* Space-Age Christianity. Morehouse-Barlow. New York, NY.
5. LEWIS, C. S. 1963. Onward Christian Spacement. Episcopalian **128**(4).
6. TELLER, E. & A. LATTER. Our Nuclear Future. Criterion. New York. NY.
7. POLLARD, W. G. 1961. Moral Effects of Nuclear Research. Paper presented at the Second Military Chaplains' Nuclear Symposium. February 28–March 2. Sandia Air Base, Albuquerque, NM.

The Moral Arguments against Military Research

BERNARD ROTH

Department of Mechanical Engineering
Stanford University
Stanford, California 94305

Insanity in individuals is something rare—but in groups, parties, nations, and epochs, it is the rule.

—FRIEDRICH NIETZSCHE

INTRODUCTION

The use of the word "moral" in the title of this paper was not of my own choice. The first I knew of it was when it appeared in my mail—carved in granite, it seemed—as part of the printed program for the conference preceding this volume. I did, some months before the conference, agree to talk as a scientist who does not do military research, and to give the arguments against doing military research. I do not recall the concept of morality entering into my conversation with the conference organizers. Personally, I would prefer not to cloak my arguments in morality. There is nothing wrong with a little morality, but to some, moral arguments represent fuzzy thinking and an impractical idealism when dealing with the pragmatic world of realpolitik; while to others, there is a militant unreasonableness associated with advocacy in the name of morality. I intend neither of these.

I do not do military research because the basic vocabulary, mentality, and objectives of a military establishment are repugnant to me. To put it simply, I just don't like it, and I am very glad I don't have to make my living that way. Obviously such pragmatism does little to enhance the discussion around the ethical issues associated with military research, so for the remainder of this paper I will summarize the arguments that I feel best speak to the case against military research.

MORALITY

For there to be moral judgments there must be an explicit or implied reference to moral law. Moral laws can be divine laws or intuitive laws based on conscience. Because interpretations of divine laws vary radically with different religious factions, ultimately morality is based on individual or collective conscience (even where the original inspiration is attributed to divine laws). Moral laws seem to have a certain mutability that usually does not exist in physical laws. The reason, of course, is that physical laws deal with ideal systems and moral laws deal with real systems and are influenced by fashions in religion and politics. Alfred North Whitehead put it the following way: "What is morality in any given time or place? It is what the majority then and there happen to like, and immorality is what they dislike."[1]

The mutability of moral arguments is evident from the different attitudes toward using consistency as a test of sincerity. For example, during World War I, the British courts ruled that conscientious objectors were "sincere objectors" if they objected to all war, and "insincere objectors" if they objected primarily to that particular war. (Sincere objectors were treated rather humanely by the courts, whereas insincere objectors received harsh penalties.) Today, however, it is unlikely that someone who objected to abortion and not the death penalty (or vice versa) would be accused of having an insincere moral objection. For the purposes of this paper I will consider any argument of conscience as a moral argument, and (unlike the British courts) I will consider all objections as sincere even if they are not universally consistent.

BACKGROUND

This paper deals with the situation in the United States as it currently exists. The present comes to us as a legacy of the Cold War aftermath of World War II. The relationship between scientists and the U.S. military was profoundly changed during the Second World War. A short first-hand summary of this was given by Vannevar Bush, the principal organizer of the U.S. scientific effort:

> When for the first time in history the decision was taken to recognize scientists as more than consultants to fighting men, I was living in Washington, and President Roosevelt called on me to head the job. Abraham Lincoln had set up the National Academy of Sciences during the Civil War, and Woodrow Wilson had authorized the National Research Council during the First World War. Both did good work, and their histories are illustrious, but neither was given large funds or authority. In the National Defense Research Committee and the Office of Scientific Research and Development in the Second World War, scientists became full and responsible partners for the first time in the conduct of war.
>
> We had, during the war, approximately thirty thousand men engaged in the numerous teams of scientists and engineers who were working on new weapons and new medicine. . . . We spent half a billion dollars. Congress gave us appropriations in lump sums and trusted us to decide on what projects to spend the money.[2]

Bush was describing a country fully mobilized to fight the most extensive war in its history. Yet at the end of the war, after a brief cutback, the scientific effort was expanded to the point where it now far exceeds that wartime level. In contrast, between World War I and World War II, the United States turned away from military activity and returned to a peacetime economy. Vannevar Bush gave the following description of the period after World War I:

> Certainly this country failed to make much progress in the application of science to military matters. Congress cut appropriations, while the Army and the Navy applied what they received principally to things other than improving their methods. Industry was not interested and took military contracts with reluctance. This reluctance was justified, for the business was unattractive, and the head of any company that developed weapons was likely to be called a merchant of death.... In this country it was not merely that people turned away from the paraphernalia for war. Civilians felt that this was a subject for attention only by military men.[2]

I like this statement, for I think it describes the proper role of a military establishment in a democratic and moral country. Regrettably, the author of this statement was lamenting and not lauding the state of affairs he was describing. Vannevar Bush, like many people today, believed that the arms race and its associated scientific research were necessary for U.S. survival. I will not go into the arguments and counterarguments about whether the arms race enhances or detracts from our security. Here I want to simply point out that the present scientific research aspects of the arms race are very much a creation of the United States, and that for the past two generations we have been one of the two leading nations in applying scientific research to the development of the implements of war.

The post-World War II arms race is different from any peacetime military buildup in history. It goes far beyond the training of troops and the manufacturing and stockpiling of weapons. It is a race without a finish line, an endless spiraling of more and more lethal, sophisticated, and costly applications of science and technology to the implements of destruction. Scientists who participate in military research fuel the arms race, and are making a political and moral statement affirming the arms race.

SCIENTISTS AND PROFESSIONAL ETHICS

Some scientists do not concern themselves with the moral and ethical implications of military work. They are either mainly concerned with what J. Robert Oppenheimer termed the scientific "sweetness" of the problems or are so consumed with the righteousness of the Cold War that they do not concern themselves with questions of means. The same is true for those who administer and promote the military–scientific establishment.

A classic example was the episode in which Hitler's rocket scientists were spirited to Alabama to be retreaded as American Cold Warriors. If these men had instead fallen into Soviet hands they would, I assume, as many of their contemporaries did, have adapted just as easily to the realities of Soviet scientific funding. In my view, these men were prototypical scientific mercenaries, and their actions, as well as those of the U.S. government's science administrators who brought them here, raise the basic issue of whether a scientist's moral or ethical responsibilities transcend the immediate sweetness of the scientific problem, the demands of the funding source, or the primacy of ends over means. There is much in this early Cold War episode that presages the melancholic nature of the current relationship between the scientist and the military.

In the universities, in private industry, and in government laboratories, there are large numbers of scientists who work for the military (either directly or by being funded from research, or research and development, contracts) but feel that taking money from military sources does not entail military work—or, if it is military work, it is benign work in that it does not have an especially military or lethal character. In fact all these scientists are aligning themselves on the side of a continued arms race; by simply putting one's head in the sand, one is not relieved of responsibility for one's actions.

The question of military funding of university research received a lot of attention during the Vietnam War and its immediate aftermath. Many scientists did not approve of the U.S. military involvement, and yet they participated in research funded by the agencies of the Department of Defense. When student activists accused the scientists of hypocrisy, many scientists, especially academics on university campuses, denied that they were collaborating with the war effort. The researchers claimed that the source of funding did not influence their work, and that military funding did not imply any special military relevance or applications of their work.

This debate took an interesting turn when the U.S. Senate passed the Mansfield Amendment to the fiscal 1970 military appropriations bill. This amendment required that the Department of Defense only fund research with a "direct and apparent relationship to a specific military function or operation." Thus the Department of Defense was required to use mission, rather than discipline, criteria in dispersing research funds. This was embarrassing to many researchers, and some continued to deny responsibility for the military aspects of their work. Some claimed not to know what the military relevancy was, while others claimed it was irrelevant, as they would do the same work even if they had a civilian sponsor.

A student-inspired, detailed study at Stanford University concluded that "the Department of Defense designs its R&D programs intelligently so that they meet projected military needs."[3] Many academic scientists continued to deny this interpretation of their research, even though military spokespeople have said essentially the same thing for years.[4] When asked by the Stanford study group to reconcile this apparent contradiction, the chief scientist of the Army Research Office is reported to have said "basic research, like beauty, is in the eye of the beholder." In other words, regardless of what the researcher saw in it the Army saw military relevance.

My own experience is that military sponsorship does influence what one researches. I have had this affirmed many times by my colleagues who do research under military sponsorship. This is not to say that the military always gets what it is paying for. Military agencies often back the wrong researchers and the wrong fields because they tend to rely heavily on an "old boy" network, but their intent is certainly to further the military rather than the civilian aspects of science. Any scientist that accepts a project sponsored by the Department of Defense is in fact a willing participant in the continued extension of the Cold War.

My experience is that many scientists lack enthusiasm for the military purposes of their funding, and question the military efficacy of their projects. For many this is a cynical business of on the one hand denying the military relevance of their work, and on the other hand working with military funding agencies to increase appropriations for scientific activities under military sponsorship. They participate in military work for career or economic advantage even though they have other alternatives.

Many scientists would prefer to be funded from sources other than the military, but given the opportunity gladly accept such funds and in this way become part of something they do not really believe in. If one is a skilled professional then there are, in this society, a variety of economic possibilities. To acquiesce out of expediency or greed into an activity one does not believe in is, in this country, generally viewed as an amoral act. I consider it an immoral act. It is one of the great failings of our educational and religious institutions that many of our brightest minds are not even aware of this issue.

Although many activities funded by the military–scientific establishment are benign, a great deal of work directly applies to weapons that are generally considered to be immoral. In the theory of war, some observers speak of "just" and "unjust" wars. They further distinguish between the justice of fighting a particular war and the justice of fighting a war in a particular way.[5] In this latter regard, activities waged against civilian nonbelligerent populations are considered unjust or immoral. Accepting this view entails asking unsettling questions about many scientific activities sponsored by the military. Atomic weapons, chemical and biological weapons, and a host of other super-killers must by their nature involve large numbers of nonbelligerents in the horrors of war.

Many justify questionable practices with the rationalization that if they did not pursue such practices someone else would. It is not ethically sound, however, to do something wrong just because someone else would do it instead.

A large number of scientists do take responsibility for participating in military research. Some would rather not do military work but find themselves without good alternatives. Many object to military work and do not do it. Then there are many who wholeheartedly approve of the arms race and willingly participate in a full range of military projects. Many of these people feel it is their duty to do this work. They enjoy it, and feel that there is nothing wrong with creating weapons to defend their country. They feel that in a democratic society such activities are necessary and proper. Let us now explore the basis for these views by considering whether the United States can claim that its military research has a special ethical or moral basis because the country has a democratic form of government.

DEMOCRATIC VALUES

Some scientists reason that a democratic country will not use its arms capriciously and unjustly. This view, however, is hardly justified by historical precedent. An early instance was the Athenian destruction of the island state of Melos in 416 B.C.; further cases of the militaries of democratic and other "civilized" states inflicting atrocities on civilian populations abound throughout history.

Any claims to a special morality underlining the government's use of the U.S. military ignore the facts. The basic Cold War operational ethic was outlined in a secret 1954 report to the White House by a commission headed by former president Herbert Hoover:

> It is now clear that we are facing an implacable enemy whose avowed objective is world domination. There are no rules in such a game. Hitherto accepted norms of human conduct do not apply. If the United States is to survive, long-standing American concepts of fair play must be reconsidered. We must learn to subvert, sabotage and destroy our enemies by more clever, more sophisticated, more effective methods than they use against us.[6]

Clearly it was Hoover's view that morality and ethics have no place in the holy crusade against communism. The Iran–Contra hearings have made it abundantly clear that segments of our government share this view and operate accordingly. The well-publicized atrocities in Vietnam and the "secret" war in Laos are examples of immoral military acts committed in the name of American democracy.

Even if the U.S. military forces did have some special moral sanctity, there would still remain the problem that the United States has become the world's second largest arms merchant. The United States now sells arms at such high volumes that it dwarfs the achievements of the legendary masters of this nefarious trade. (I have in mind Britain's Sir Basil Zaharoff and Germany's Krupps.) The United States exports billions of dollars worth of military equipment annually (the World Almanac puts the 1985 figure at $12.3 billion) and has no real control over how these arms are used. The arms trade profits from military research, and military research profits from the arms trade. Scientists who do not want to contribute to this trade should not contribute to military research.

The post-World War II history of U.S. arms development, deployment, and sales is full of questionable, and I would say immoral, actions. For example, what moral purpose was achieved in selling the Shah of Iran $18 billion worth of arms? The recent highly publicized, and probably illegal, covert supplying of missiles to Iran is simply more publicized than, but not different in kind from, many similar arms transfers. In fact, how can conscientious scientists ever know the extent of U.S. involvement and what atrocities have been facilitated because of their work? (It is estimated that President Reagan alone authorized over 50 major covert operations and signed over 200 secret National Security Directives during his first seven years in office.[6])

The question is: are the scientists who create the arms and who derive their livelihoods from their sale culpable for their use? I believe they are in some measure responsible, for they must surely know that the results of their labors are to be used to further the quest for U.S. hegemony and not for the righteous defense of their homes and liberty.

Even if it would be possible to make moral claims for the military use of scientific developments, there still would be ethical problems for participating scientists because the very basis of military activity tends to be antithetical to the principles of science. The military lies to the citizens of its own country and conceals from the public what it is doing. So military work tends to corrupt one of the basic tenets of science, the open and collegial sharing of knowledge for the common good. Scientists who work for the military lose control of their freedom to communicate freely with their colleagues.

CONTROL OF RESEARCH

The scientific community often refers to the early beneficence shown by the Office of Naval Research in supporting the scientific enterprise in this country without specific emphasis on military matters. This brings up the question as to why, in a democracy, the military is one of the main conduits for the government's support of science. At this time, is not the National Science Foundation the proper agency for the general advancement of science? Should not the National Institutes of Health, the Department of Agriculture, and the Department of Commerce be concerned with the promotion of the basic and applied sciences with relevance to their areas of responsibility?

Why is it that the military is funded as though it were the U.S. equivalent of MITI (the Ministry of International Trade and Industry, the Japanese agency that coordinates national scientific initiatives)? What is the rationale for channeling large amounts of financial support to science through the military? Why should the military get to decide which fields get support? Why should the military support fields with only tenuous military relevance?

With the military establishment as large as it is, almost everything is of some concern to it. The military is the largest single consumer in various nominally civilian sectors: housing, medicine, agriculture, manufacturing, and transportation are all of concern to the military. So what we have is a state within a state, and in that sense every endeavor has some military relevance. The issue then becomes what is a proper military area and what should be left to civilian agencies. By and large the military receives its funds from the Congress for direct military purposes. By assisting the military to avoid narrow interpretations of its mission, the scientific community is in collusion to mislead and defraud the Congress and the American people.

There is a feeling among scientists that Congress will not support a large scientific enterprise on its own merits. They reason that the most expedient way to get Congress to appropriate large amounts of public funds to scientists is on military grounds or the need to solve specific problems in health, agriculture, or manufacturing. The trends have certainly been in this direction. In 1980, the federal research and development funds were divided approximately equally between the military and civilian funding agencies; by 1987, the military received twice the research and development funds the civilian agencies received. Thus, branches of pure science that seem to have some potential military relevance get a bigger slice of the federal pie.

Even supposedly civilian funding agencies such as the Department of Energy and the National Aeronautics and Space Administration have strong military biases in their funding. A classic example has been the relatively high funding levels that Congress has long appropriated—mainly through the Department of Energy and, earlier, through the Atomic Energy Commission—for the study of fundamental particles. The financial success of this field was largely the result of atomic physics having produced the atom bomb during World War II, and of later generations of physicists having worked on more sophisticated nuclear weapons. There is always the hope that each new generation will also come up with "something useful," even while working on fundamental problems. Of course it does not hurt a field to have the Soviets active in it; then funding levels can be maintained on the grounds of avoiding what is called technological surprise.

Most money for laser research, computer science, and scores of other areas now comes from military sources, even though, just as in the case of particle physics, many of the scientists believe that the military relevance of their work does not, in fact, justify military support. Once civilian institutions become dependent on military funding of the scientific enterprise, they become part of the lobbying effort to increase military funding. In addition to misleading the Congress, the universities and research laboratories are perverting the scientific enterprise by becoming part of the pressure group for larger military budgets.

I feel it was wrong for the President of the Association of American Universities to have testified on behalf of 50 research universities, before the House of Representatives Committee on Appropriations, in favor of appropriations for the Department of Defense. His testimony bordered on the obscene when he blatantly laid out the strategy for the Faustian bargain. Give the universities money, and they will deflect students away from other scientific work and "hook" them into military work:

> I think it . . . would be helpful to the Department of Defense, to enlist the loyalty of a group of students with funds that it awards, to enable students to pursue their graduate studies with the sense that they are beginning an engagement in work that is of interest, that will ultimately be of interest to the Department of Defense and related agencies.
>
> So I think from that point of view, to the extent that the Defense Department supports graduate students in the sciences and engineering, it is beginning to build

a cadre of scientists and engineers who will be participants in its programs in the future. . . .

. . . If they are engaged early in work that is intellectually stimulating to them and that has some promise for the future and is supported by the DOD, it seems to me you are well on the way to having them hooked into that enterprise for a long time.[7]

To me this testimony advocates something akin to the neighborhood drug pusher "hooking" the young school children with free samples. Here the "free" hit is a graduate stipend, then the student is hooked to a lifetime of work for the Department of Defense. The military has long had a special interest in the universities.[8,9] This testimony is a vivid example of the collusion of university administrations in pursuit of funds, and the military in pursuit of university scientists.

As the testimony indicates, judicious use of research funds and a cooperative university research establishment provide the military with de facto control over the work of individual faculty, students, entire departments, and even entire fields. Other than for a short period during the Vietnam War, the Department of Defense has been very successful at leveraging the money spent at premier institutions into gaining control over large segments of the educational enterprise.[8,9]

This testimony points to another problem with military research: for many areas military funds provide most of the employment. So if students are drawn into certain areas during their university training, they may not find any employment opportunities in the civilian sector. In this way students in many areas of physics and engineering are essentially forced into military work after graduation. In a short time their skills become so specialized and particularized to the military enterprise that they must either continue in military research or leave their profession.

MORALITY AND ECONOMICS

In one way or another large military establishments keep their own nations in bondage. In certain military dictatorships it is obvious that the size of the military is not commensurate with any real foreign threat. The military's primary function is as a tool to enforce domestic policy on a reluctant citizenry. In other countries a large military is a force in domestic politics in more indirect ways. For example, in the United States and the Soviet Union an implied mutual threat is used by their military-industrial-scientific establishments to persuade their nations to make colossal economic expenditures in support of their military establishments. One of the basic arguments against participating in scientific research for the military has to do with not wanting to associate with this depletion of national resources for military purposes.

The spectacular economic success of post-World War II Japan is a graphic example of what a nation can accomplish once it is free of the tremendous economic burden of an oversized military establishment. Japan spends only a tiny fraction of its gross national product on its military, and has an educational and training system focused on civilian technology. Virtually all of its scientists and engineers find employment in the civilian sector. It is these aspects of their society, rather than their much-publicized robots, that should receive a large share of the credit for their economic miracle.

The use of scientific resources for military rather than social purposes is morally wrong. Given the social ills of our society and the poor condition of the infrastructure in many places in America, and also the desperate poverty of much of the world, the squandering of talent and money on the junk of war is wrong.

ARMAGEDDON

Much current debate rages about specific types of scientific activity: nuclear weapons development; chemical and biological weapons; and, most recently, the Strategic Defense Initiative ("Star Wars"). Each of these topics has become a cause celèbre in the political battles over the proper role of the scientist in the Cold War. Much has already been written about these topics, so I restrict myself here to a few comments in order to devote my limited space allotment to less well discussed arguments against military research.

Nuclear arms are the ultimate symbol of the folly of the arms race. In his landmark book, Jonathan Schell[10] fully describes the catastrophe that these weapons can bring to our entire planet. Clearly these weapons exist only because of scientific research; without the scientists' willing and enthusiastic cooperation, these weapons would not exist (and the world would be a much safer place). The history of the development of the first bomb has been told in many places. What is interesting for our present purposes is to realize that the scientists who were instrumental in creating the bomb almost immediately lost any semblance of control over the product of their immense and brilliant efforts.

Two interesting cases in point are those of Niels Bohr and J. Robert Oppenheimer. Niels Bohr, the famous Danish physicist, was one of the first to understand that the long-term dangers of the atomic bomb were far greater than any immediate advantages. He communicated his concerns to both Roosevelt and Churchill, and, even though he had been helpful in alerting Americans to early important experiments, he was subsequently treated as a security risk and totally ignored.

Oppenheimer, the leader of the team that developed the atomic bomb, ran into trouble when he opposed the development of the hydrogen bomb. He headed an advisory panel whose members unanimously decided that the development of the hydrogen bomb was not warranted. One of their most important reasons was their belief that the moral standing of the United States would be diminished if we developed such a weapon. Hard-liners persuaded

President Truman to ignore the panel's recommendation and to authorize the weapon's development. Shortly afterward Oppenheimer was stripped of his security clearance and removed from his government posts.

These cases vividly illustrate that ultimately scientists have no control over the results of their work, and that this is equally true for even the most prestigious and famous. The only way to control the consequences of one's work is to control what one works on in the first place. Not even this will assure complete control, but it is certainly better not to create something that by its very nature is contrary to one's principles.

Clearly biological and chemical weapons come under the heading of something that scientists should refuse to work on. Most medical researchers think of their science as being applied to disease prevention and the alleviation of human suffering, so very few of these scientists would chose to work on such weapons. To counteract this the military mislabels all such work as defensive, and uses subtle methods of recruitment.

In a recent article, Professor King described how the biology department at the Massachusetts Institute of Technology was coerced into accepting Office of Naval Research funding for a new biotechnology program:

> Funding of biological research by the military serves several purposes. It contributes to the increasing incorporation of the university into the military–industrial complex. It provides a veneer of respectability to cover the support of the military for its more destructive projects. It increasingly focuses academic research on problems of concern to the military. And it provides direct and indirect support for the resurgence of BW (Biological Warfare) research.[11]

The intrusion of the military into academic biology follows essentially the same tried and true techniques that have worked well for the military in the physical sciences and engineering.

Star Wars research has been the subject of one of the largest protests ever launched against a military undertaking by U.S. scientists. Although the reasons for the opposition were varied, there were two almost universal elements: a rejection of the hype associated with the promise of a leak-proof defense, and the desire of scientists not to be used by military interests to enhance the prestige of a dubious undertaking. Those scientists who joined in expressing their concern and boycotting the project displayed a degree of independence and ethical behavior heretofore remarkably absent from the scientific community. One should hope that their actions have dampened congressional enthusiasm, and will ultimately discourage a costly and dangerous acceleration of the arms race.

PROPER MILITARY RESEARCH

What is a proper role for military research? To date, most military research has been directed toward continuing and accelerating the arms race. Throughout the Cold War, research aimed at perpetuating the arms race has been much more sought after than knowledge to end it. This is a subversion of our humanistic values, and the acquiescence of the scientific community to this is immoral. There have been some who have raised these issues. Interestingly, a goodly number of these are individuals who were part of the defense establishment and, because of conscience, decided to speak out against the current state of military involvement in our society (former weapons scientists play prominent roles in the *Bulletin of the Atomic Scientists*, the Federation of American Scientists, and the Union of Concerned Scientists).[12]

There are, in some cases, distinctions between defensive and offensive weapons. Some feel the defense of rights is a reason for fighting a war, and for these individuals work on defensive weapons would be an act of conscience. If their belief rules out every other sort of war, then the distinction between offensive and defensive weapons becomes important. This is sometimes a very subtle distinction, and it is often exploited by the military: for the sake of public relations offensive weapons are developed under the guise of defensive ones.

For others, arms reduction and stability of the status quo are important. For such individuals research in arms reduction and treaty verification seems appropriate. Scientists have at times played a vital role in this area. Their opposition to atmospheric testing led to the Limited Test Ban Treaty, and their opposition to antiballistic missiles was probably instrumental in regard to the 1972 Antiballistic Missile Treaty.

Regardless of which path a scientist chooses, it is important to develop areas of scientific enquiry that do not limit the scientist's role to weapons technician and supporter of the arms race. Scientists in democratic countries have a special duty to speak out. For them the penalties are at most economic and loss of peer-group approval. The personal risks are far less than those for our colleagues in more restricted political environments. If scientists in the United States were to distance themselves from the military, they would be setting a good example for colleagues in other countries. It is only proper that the scientific community in the country that first built an atomic bomb take the initiative in halting the wasteful and potentially disastrous arms race that the bomb helped to fuel.

The moral choice is spelled out by Albert Camus when he advises us to become "neither victims nor executioners" and to make a conscious choice as to which side of the "terrible dividing line" we choose to stand on:

> All I ask is that, in the midst of a murderous world, we agree to reflect on murder and to make a choice. After that, we can distinguish those who accept the consequences of being murderers themselves or the accomplices of murderers, and those who refuse to do so with all their force and being.[13]

As individuals, as scientists, and as a nation, we must examine our consciences and make sure we are on the moral side of this line.

SUMMARY

Military research has become part of the established order in scientific inquiry throughout the world. Because of the awesome power of the military arsenals and the potential for ending or altering many forms of life, the questions associated with military research cannot be regarded as purely scientific or political. More than in any previous period of history, the issues associated with military research have become a moral problem. It is argued that the moral responsibilities of scientists are not altered because a country has a more democratic form of government. In fact, scientists with the most personal freedom should be the ones pressing the moral arguments against military research. It is further argued that research tends to follow paths pointed out by authorities, and when that authority is the military the research will always have unwanted side effects that are immoral and counterproductive for the scientific enterprise and the well-being of the citizenry of that country. The problems raised by military research bring to the fore basic issues associated with the appropriate size and funding of a country's scientific enterprise, and the question of how best to achieve the goals associated with that enterprise.

REFERENCES

1. WHITEHEAD, A. N. 1954. Dialogues of Alfred North Whitehead (as recorded by Lucien Price, August 30, 1941). Little, Brown. Boston, MA.
2. BUSH, V. 1968. Modern Arms and Free Men. Paperback edit. MIT Press. Cambridge, MA.
3. GLANTZ, S. A. & N. V. ALBERS. 1974. Department of Defense R&D in the university. Science 186: 706–711.
4. MACARTHUR, D. M. 1970. Defense technology: Benefits to industrial progress. Def. Ind. Bull. 6: 1–3.
5. WALZER, M. 1977. Just and Unjust Wars. Basic Books. New York, NY.
6. MOYERS, B. 1987. The Secret Government: The Constitution in Crisis. Public Affairs Television. New York, NY.
7. ROSENZWEIG, R. 1985. Congressional Testimony on May 15, on the Department of Defense Appropriations for 1986. Committee on Appropriations, House of Representatives, 99th Congress, First Session; Subcommittee on the Department of Defense. Part 8: 771–966. U.S. Government Printing Office. Washington, DC.
8. ROTH, B. 1986. The impact of the arms race on the creation and utilization of knowledge. *In* The Optimum Utilization of Knowledge. K. Boulding & L. Senesh, Eds.: 295–313. Boulder Press. Boulder, CO.
9. NELKIN, D. 1972. The University and Military Research. Cornell University Press. Ithaca, NY.
10. SCHELL, J. 1982. The Fate of the Earth. Alfred A. Knopf. New York, NY.
11. KING, J. 1988. Biology goes to war. Sci. People 20(1): 17–20.
12. EVERETT, M. 1989. Breaking Ranks. New Society. Philadelphia, PA.
13. CAMUS, A. 1986. Neither Victims Nor Executioners. New Society. Philadelphia, PA.

Masters of War
The Moral Arguments and the Traditions of Ethics

DOUGLAS MacLEAN

Department of Philosophy
University of Maryland Baltimore County
Catonsville, Maryland 21228

INTRODUCTION

I have been asked to comment, from the perspective of someone trained in philosophy, on the debate about the ethics of military research. This debate is carried on by engineers who know firsthand the nature of research primarily oriented toward military purposes. They know intimately some of the people who are involved in military research; they know some people who have chosen not to become involved and who have subsequently paid a cost for this choice; and they know the culture of the universities, laboratories, and private firms in which military research is conducted. Their professions, like my own, must surely contain a share of rogues, people with warped sensibilities that fit evil or demented stereotypes. But they must also contain a far larger number of sensitive professionals, committed to doing good. And a large segment of this research community probably goes to work each day and does the job at hand without thinking much at all about the ethics of their enterprise. I suspect that very few of them would fit the image of the "masters of war . . . who build to destroy."

An adequate account of the ethics of military research would no doubt have to be based on a thorough knowledge of the culture in which it is carried out and the people who do it. We would have to know something about the choices they have to make and the reasons for them, and we would have to examine the choices they choose not to face and the explanations for them. I enter this discussion with pretty close to complete ignorance of this background. Having delivered this caveat, I want now to defend a strong thesis. Given my ignorance of the details, my argument is a tentative one.

A STRUCTURAL PROBLEM

My thesis is that, except for the normal expected cases of fraud and malfeasance, viciousness and spite, arrogance and deception, all of which we would find in any profession, what bothers us about the ethics of military research has little to do with individual character and motives. What bothers us is, rather, a structural problem. It has to do with the arrangements of a system and the incentives it creates. To the extent that the system is morally condemnable, the culprits are much harder to identify because the responsibility is more broadly distributed. But I am less interested in assigning responsibility here than in analyzing the problem.

Let us begin by asking the following question: When we make a moral judgment, when we praise or condemn, what is the object of that judgment? Clearly the objects of moral judgment are many in kind, and the nature of judgments varies accordingly.

For example, we judge *outcomes* or, more generally, states of affairs, which we frequently think of, for example, as good or bad. Thus, we morally judge outcomes on the basis of the amount of pain and suffering they contain, or on the basis of the amount of happiness or the fairness or equality of the distribution of benefits and burdens.

We also judge *people*. We offer moral praise for their virtues, such as generosity, integrity, sensitivity, dedication, and so on; and we condemn people for their vices. We judge characters explicitly (although sometimes this tendency to judge is itself considered a vice!) and implicitly, as when we forgive an action that has bad consequences because we impute good motives to the actor.

Finally, as just suggested, we judge *actions*. For these judgments, we sometimes apply the vocabulary typically used to judge character and sometimes the vocabulary typically used for states of affairs. The semantics of moral judgments do not divide crisply. But we have a vocabulary that we typically apply to actions, too, and this is seen, for example, when we designate actions as being right or wrong, thoughtful (although this can be a judgment of character too), or horrible (although a state of affairs can also be horrible).

These are three different objects to which we commonly apply moral judgments. I do not mean to suggest that they are exhaustive, but for the purposes of my argument it will be sufficient to discuss these three. Obviously, they are closely related and dependent on each other in many ways. We often judge an action by its outcomes, and according to some moral theories, the judgment of actions is derivative on the assessment of outcomes or states of affairs. Our common sense morality, however, is not so clearly regimented. We may ultimately want to revise our common sense morality, but this morality tends also to judge some actions independently of outcomes. We govern our actions in part by moral rules, and we apply these rules directly to our actions, even when their expected outcomes are clearly not always beneficial. Some moral theories claim that these rules and the judgment of actions are morally fundamental, but again this kind of regimentation does not square well with our common sense morality.

Actions are (typically) performed by people, and they are performed out of different motives. Motives, too, affect our moral assessment of actions. The assessment of actions and judgments of character are thus interrelated.

But again, we judge persons or characters partly on the basis of the actions they tend to perform, but partly on aspects of their personality that we glean in ways independent of a moral tabulation of their actions. And again, some moral theorists attempt to argue that issues of virtue and vice are fundamental to morality, but our common sense morality does not fully conform with these theories either.

Normally we do not think about the different objects of moral evaluation and how they are to be distinguished or related. One reason we do not is that the objects of moral evaluation tend in the vast majority of cases to hang together in a conceptually convenient way. Virtuous people tend to perform morally right actions, and right actions tend to produce good states of affairs. People with vicious characters must do things we find wrong, and these wrongful actions will usually have bad consequences. It may thus seem not important to distinguish the objects of moral evaluation carefully. Even if we acknowledge moral judgments ranging over different objects of moral evaluation as being logically distinct, it would be inconceivable that they would not correlate closely.

An example will help to illustrate this point. A kind of action might be thought to be right because it conforms to a moral rule, and we tend to justify moral rules in ways that do not necessarily appeal to the consequences of performing this kind of action. It is good to be honest and keep promises, and we endorse a rule of promise-keeping not necessarily because we enjoy the consequences of being able to rely on what others say and thus coordinate our activities, but because being truthful shows respect for other people. It is part of what is involved in treating other people properly. Nevertheless, it would be inconceivable that we would accept the moral rule to keep promises if keeping promises tended in most cases to cause pain or to produce bad consequences. We would find it hard to imagine a world where that was the case.

MORAL COHERENCE

I will refer to this claim about how the moral world hangs together as the principle of moral coherence. It is not clear whether this is an empirical principle, one that reflects a deep fact about our world, or a conceptual principle, that is, some presupposition about morality and the semantics of moral language. But the principle of moral coherence explains why we do not usually need to distinguish between the objects of moral evaluation and the different kinds of moral judgments appropriate to these objects.

There are, nevertheless, important situations in which the principle of coherence can fail. One of these, not too far removed from the subject at hand, is in the context of deterrence.[1] Here it is possible that making and communicating an intention to commit an act that would, if performed, be morally wrong, can produce good consequences. Deterrence works, roughly, by making conditional promises one hopes will never have to be fulfilled. Worse still, in some situations an effective deterrent might depend on the threat (or the conditional promise) being credible, and it may require an individual who lacks and is known to lack certain virtues to be able to make such a credible threat. A moral saint might be unable to deter actions that produce bad states of affairs that less saintly people (or perhaps even totally crazed people) could succeed in deterring.

Situations where the principle of moral coherence fails might be rare, but they may also be important. And it is this failure, the way that evaluations of people, actions, and states of affairs come apart, that leads us to regard these situations as paradoxical. In these situations, our common sense morality often fails us. We do not know when, or under what conditions, to prefer virtuous actors, right acts, or good consequences.

Now, to bring the point home, we also find failures of moral coherence in situations of role morality, and this is why role morality, like deterrence, creates important problems for moral theorists. The troubling feature of role morality is that certain social institutions seem to be morally justified (or even required) because of the tendency of the workings of those institutions to bring about good results, or results that are better than the states of affairs we could expect in the absence of such institutions. But individuals working within these institutions might be required to act in ways that could not be justified outside of the roles actors may assume within these institutions. The world may be such that, unfortunately, we need as nations to engage in espionage. It may be better to be in a world in which we spy on each other than to be in a world in which nations remained more ignorant than they are of the military strategies and preparedness of their adversaries. But those who are spies often engage in actions that would not be morally permitted if they were not acting within institutionally sanctioned roles, and they may be forced by these roles to lead secret or dishonest lives that corrupt their characters in important ways.

More commonly, it is only physicians acting within their roles as medical professionals that we allow to incapacitate people or to slit them open with a knife. And we often remark on the paradox of lawyers serving the noble ends of justice by withholding what they know to be true about their clients, by badgering innocent people on the witness stand, and so on. We justify certain institutions or professions by their tendency to produce good states of affairs, but the professions define the roles that individuals must perform. Paradoxes may arise when these roles call for people to do what would not be permitted of them outside these roles, and we can worry about whether the effects of accepting a role might corrupt a person's character, or what the limits are to which a person might go within a role.[2] Is it a good lawyer or a bad one who exploits procedural technicalities to acquit a client he or she knows to be bad? Is it a good physician or a bad one who decides to forgo treatment in a hopeless case so as to "protect" the family from having all the information that would force them to make or endorse the tragic choice?

INDIVIDUAL, PROFESSIONAL, AND STRUCTURAL CONSIDERATIONS

In considering the ethics of military research we need to focus on the structure of the entire defense establishment, and especially on the roles of those researchers who work directly or indirectly within it and the natural incentives that form an important part of these roles. We should see the bombmaker, eagerly and enthusiastically dreaming up ever more deadly weapons, in this context. We should, that is, see this person as bringing enthusiasm and dedication to his work, traits that we would normally praise. Likewise, we should be reluctant to condemn the engineer in a national laboratory whose confidence about the prospects for the Strategic Defense Initiative seems to outstrip that of a lot of other experts. After all, isn't it to be desired, generally speaking, that engineers assigned to design new products have a robust optimism about their prospects? The moral problems here may lie more with the institutions and the way they work than with these professionals.

Similarly, there are powerful economic and professional incentives for scientists and engineers to dedicate their energies where the work is exciting and the funding plentiful. I suspect it is often wrong to demand too great a moral sensibility on the part of each of these individuals in examining each project they undertake. And it is probably too much to ask them to forgo jobs and grants in their fields in order to avoid dirtying their hands in weapons work. Besides, it is not this area of research generally that we deplore — there is close to a consensus, I believe, that national security is morally justified, or even required — but particular projects and, more importantly, prevailing incentives within the military–industrial–research complex.

I do not mean to suggest that scientists and engineers at our laboratories and within the weapons factories are beyond the range of moral judgment. Far from it. In seeking to resolve our moral dilemmas, we place a burden on these scientists and engineers because we often demand that they help us determine whether some highly technical undertaking truly promotes the cause of peace and security or not. So we do expect them to be able to rise above their occupational roles, and we include these sometimes conflicting obligations in our conception of their professional responsibilities. The engineer who blows the whistle on some lucrative but foolish enterprise is a noble person, and the ones who let loyalty override good sense are condemned and occasionally punished.

But I think we will progress very little if this is where our moral investigation comes primarily to rest. We must instead open it up to look at the entire defense establishment — the military–industrial–research complex. Until we bring the executive, the politician, the banker, and investor into this discussion, until we come to terms with the entire incentive structure that promotes waste (that is, promotes projects independently of their value in preserving and stabilizing peaceful relations between adversaries), until we figure out how to create incentive structures that will enable scientists and engineers to combine loyalty and dedication with an ability to be openly critical, without destroying morale or requiring great personal sacrifice, we will be focusing our moral scrutiny both inefficiently and, I believe, unfairly.

REFERENCES

1. These issues are discussed in detail in KAVKA, G. 1978. Some paradoxes of deterrence. J. Philos. **75**: 285–302.
2. For further discussion, see especially NAGEL, T. & M. WALZER. 1974. *In* War and Moral Responsibility. M. Cohen *et al.*, Eds. Princeton University Press. Princeton, NJ. Also see LUBAN, D. 1988. Lawyers and Justice. Princeton University Press. Princeton, NJ.

The Moral Arguments and the Traditions of Religious Ethics

ROGER L. SHINN

Reinhold Niebuhr Professor Emeritus of Social Ethics
Union Theological Seminary
New York, New York 10027

What have religious traditions to do with military research and action? The relation is both important and problematic. To ignore it is foolish. For many people, ethics is an integral part of religious belief. Even the nonreligious or irreligious must recognize that some people and societies, probably a majority of people around the world, understand their ethics as part of their religion. A glance at the world today quickly shows the importance of religious traditions, for better and worse, in international conflicts. But the problems in the relationship are immense. Both the significance and the confusions of the relationship rise out of the nature of religious ethics.

Religious morality involves claims with a transcendent dimension. The transcendent dimension is conceptualized in many ways. It may refer to a God, who is creator and judge of humankind. It may affirm a spiritual realm that infuses the natural. It may describe a universal loyalty that encompasses particularistic loyalties of individuals, tribes, or nations. In one way or another, a religious morality completes and often overrides common, prudential considerations.

That transcendent reference is frequently corrupted by the claim of transcendent and universal justification for partisan demands. Rather than subordinate such demands to a higher judgment, individuals or groups often identify their own interests with transcendent norms. This habit is known in the biblical tradition as idolatry.

TWO DIFFICULTIES

Diversity

The first great problem is the ethical diversity of religious traditions. Religions nurture fanaticism and reconciliation, revenge and forgiveness, war and peace. In relation to the subject of our consultation, they say almost everything—which means almost nothing.

But that is not the end of the matter. There are, among all the diversities, evidences of significant convergence.

In the spring of 1987, a week-long seminar on "The Future of Mankind and Cooperation among Religions" gathered in Tokyo. Its sponsors were the United Nations University, the Japan Foundation for the United Nations University, and the Japanese Committee of the World Conference on Religion and Peace. Its participants came from the main religious traditions of East and West. They did not seek to produce a manifesto. But they demonstrated that believers from many traditions, without diluting their specific beliefs, can cooperate toward many social ends. In particular, they can together recognize the unprecedented peril of nuclear war, and they can agree on many shared efforts toward the achievement of peace.

I must not exaggerate that consensus. People who attend such seminars are those who already appreciate traditions other than their own. Some of them were intensely aware that their cobelievers—in Northern Ireland, in Lebanon, in Israel and the surrounding lands, on the border of Iran and Iraq, in southern Africa, and in the United States—were using their beliefs to inflame hostility. But that chaos of belief does not nullify the convergence. My statement here is that of one who, living in the Christian tradition, seeks to enhance that convergence.

Conviction and Reality

A second problem comes in the relating of religious commitments to political reality. Religious conviction is inherently fervent. It asks for unqualified loyalties and commitments. It is likely to be unyielding. It tends to see compromise as temptation, maybe betrayal. Sometimes its idealism, in the face of the world's stubbornness, leads to a utopianism that is irrelevant to reality. All these qualities relate uneasily to the negotiation, the give-and-take, the trade-offs of political situations.

To these perennial difficulties add the unprecedented character of the nuclear era. Rarely can religious traditions, of themselves, prescribe public policies in unprecedented situations. Religions must interact with secular disciplines. Religious faith can contribute to the body politic by awakening ethical sensitivity, but it is rarely competent to prescribe policies without the help of skills in the political arts.

The interaction of faith and secular skills has a long history in the development of religious ethics of politics and economics. Now, increasingly, the interaction must develop with science and technology. A scientifically uninformed religion cannot address wisely the ethical issues that rise out of enhanced human powers unknown to the past.

Yet the contribution of religion is critically important. Science as science does not determine the good. The ethical uses of science are not scientific decisions. The direction of scientific research is not exclusively a scientific question; it is a human, a political, and an ethical question.

Therefore I do not expect the traditions of religious ethics to provide precision in judgments on issues of public policy, especially issues that rise in our high-tech world. What the traditions provide are ethical insights and an ethical context within which persons and societies can appropriate and evaluate scientific and technological activity. Religion cannot impose those insights and that context. It can offer them and seek to make them persuasive.

FOUR CONTRIBUTIONS FROM THE RELIGIOUS TRADITIONS

Initial Assessment of Military Research

In the convergence of religious traditions, which I have already mentioned, peace is recognized as a good—a very great good, although not the only good. Conflict is part of life. Conflict need not always be violent, and some highly committed believers—for example, Mahatma Gandhi in the Hindu tradition and Martin Luther King, Jr. in the Christian tradition—seek to avoid all violence, particularly war. The more common belief is that war is sometimes justifiable as defense against aggression or as revolution against intolerable tyranny. Out of that belief has come the concept of the "just war"—or, because wars are never purely just, the "justifiable" war. The historical record shows many detailed attempts, both religious and secular, to define criteria for the ethics of war. These criteria vary in detail, but fall into two general categories governing 1) the purpose of the war and 2) the methods of fighting it. Because all combatants usually claim justice for their purposes, else there would be no war, international agreements are more frequent on the ethics of methods.

Modern technologies make the old restraints hard to apply. For example, there is sometimes a clear distinction between the killing of invading soldiers and the indiscriminate destruction of cities with their hospitals, schools, and populations. But traditional distinctions between combatants and noncombatants are increasingly obscure. In industrialized societies the manufacturers of tanks, bombers, and high explosives have as much military potency as uniformed soldiers. But when factories become targets, distinctions between combatants and noncombatants fade. In the area bombings of World War II, the old restraints eroded. In an all-out nuclear war they would disappear. Ethical thought about nuclear war centers not on the appropriate use of nuclear weapons but on the avoidance of such war.

Even the most reluctant advocacy or acceptance of war implies, at least in the modern world, an acceptance of some scientific military research. A society cannot ask its soldiers to fight while denying them weapons. Even so, there are some ethical restraints on weaponry that imply restraints on some research. One belief, widely held, is that some forms of warfare—for example, biological, chemical, and nuclear—are so horrendous as to permit no ethical validation.

The Biological Weapons Convention of 1972 (still unratified by many of its 122 signatories) has been described as "the only disarmament agreement in history" requiring governments "to forego the possession, production, as well as the use of, a class of weapons." Its goal is "to exclude completely the possibility of bacteriological (biological) agents and toxins being used as weapons."[1] One reason for the moral revulsion against biological warfare is a kind of primal belief that biomedical research has the purpose of protecting rather than destroying life. A more pragmatic reason is that living organisms, reproducing themselves, can spread through a population, far beyond the boundaries of their intended targets. Conceivably they could become weapons for terrorists. A panel at the 1989 meeting of the American Association for the Advancement of Science warned against a biological arms race.[2] Several hundred biologists and chemists have signed a pledge "not to engage knowingly in research and teaching that will further the development of chemical and biological warfare agents."[3]

Such a statement lacks precision because basic research has unpredictable outcomes. The same research may start a chain of discoveries leading both to healing and to destruction. When I once consulted a scientist who had done important government work, he estimated that three-quarters of what had gone on at Fort Detrick, as research on bacteriological warfare, was indistinguishable from work at the National Institutes of Health. But the remaining one-quarter is important. Furthermore, the pledge has important public, symbolic meaning.

Public concern has been most evident in the case of nuclear weapons. The resultant controversies express both ethical passion and ethical perplexity. It is easy to oppose nuclear war, harder to formulate precise policies about research. Religious communities have, for the most part, supported the Limited Test Ban Treaty, enacted in 1963. Here was a history-making agreement that restrained specifiable kinds of military research. It was possible because tests in the atmosphere did obvious harm and because compliance with the treaty was verifiable by technical means. The United States and the Soviet Union came close to agreement on a comprehensive test ban, but failed, ostensibly or actually because of problems of verification.

Today, if some military research heightens the destructive possibilities of war, other research may be beneficial. The research that led to the observation satellite (the spy in the sky) made possible the mutual verifiability assumed in the U.S.-Soviet agreement on intermediate-range nuclear forces in Europe. Again, the case can be made that a shift from a missile that carries several independently targeted warheads, such as the MX, to a missile that carries a single warhead, such as the midget-man, would reduce the temptation of adversary nations to launch a first strike. Such a shift in weapons would require some further research.

Religious ethics, in general, regards defensive weapons as preferable to offensive weapons. If there is no absolute distinction between the two, there are some functional distinctions. Both Ronald Reagan and Mikhail Gorbachev have publicly advocated a shift from offensive to defensive weapons. The sincerity and accuracy of their statements requires investigation. But their agreement at least suggests that ethical convictions might influence the direction of research.

I have been making the case that religious communities, unless explicitly pacifist, cannot register total opposition to military research. They may well advocate restraints on some types and directions of research.

Social Impact of Military Research

A second kind of ethical restraint concerns not the specific projects of military research but its overwhelming quantity. Concentration of research on military projects skews the scientific–technological enterprise, the economy, and the ideological climate of nations and the world. Granted, military research has some by-products that benefit society in general; it also diverts energy from basic human needs.

Harrison Brown reckoned a decade ago that some 375,000 scientists and engineers, "perhaps 40% of the world's total pool of highly qualified research people," are "dedicating their research skills to military problems."[4] Some economists attribute Japan's economic miracle to its abstention from large-scale military production. Other industrial nations, diverting much of their scientific skill and energy to military projects, have trouble competing with the Japanese in production of automobiles and computers.

Suspicions are mounting that the military establishment is inept in sorting out even its own priorities. According to press reports, Air Force specialist Tom Amlie has shown that the military bureaucracy since the Civil War has usually made bad choices in technical innovations.[5] This is a serious failing when a new bomber costs $520 million per plane and, by some evidences, is crash-prone. It is often said of some officials in the Pentagon that "they never saw a weapons system they didn't like." Such officials, apparently incompetent in choosing their own priorities, are now choosing technological and economic priorities for the whole society.

A few years ago I learned, in conversations at the Massachusetts Institute of Technology, that some students, contrary to their personal and ethical preferences, are drawn into military research because that's where the money and the action are. I do not mean that financial greed lures these students. It's just that scientific research is costly, and research follows the money.

The justification of some military research is not an endorsement of the vast quantity of such research in today's world. That, I regret to say, is neither a dramatic nor a precise statement. "Some, but not too much" is not the language of a thundering manifesto. It may be more honest than a pretentious pronouncement.

Human Fallibility and Sin

A third religious insight is the sensitivity to human fallibility and sin, with the consequent distrust of systems that can and often do go wrong. It is foolish to amass great destructive force with the proud confidence that "our" side is so pure and infallible that we can never misuse it.

China is currently denouncing many cruelties of the Maoist era. The Soviet Union is denouncing the blunders and wrongs of Stalinism. The United States has generally recognized the misadventures of its recent past. The public knows that both the United States and the Soviet Union have shot down passenger planes in error. The technologies are in place to multiply these errors on a colossal scale.

Soon after World War II the British historian, Herbert Butterfield, in a book called *Christianity and History*, wrote: "The hardest strokes of heaven fall in history upon those who imagine that they can control things in a sovereign manner, as though they were kings of the earth, playing Providence not only for themselves but for the far future—reaching out into the future with the wrong kind of far-sightedness, and gambling on a lot of risky calculations in which there must never be a single mistake."[6] The years since have added force to his argument.

Security and Risk

A final insight can be stated briefly, but is as important as any of the others. It is the understanding, almost universal in religions, that no human contrivance can guarantee absolute security. Life is a precious gift and civilization an achievement: both are worth cherishing and protecting. We rightly try to secure them against obvious perils, including military aggression. Throughout much of history the enemy in war has been a hostile army. In these latter days, we have moved into an era when war itself is a greater enemy than any national adversary.

Prudent military defenses may still deter war. But any pretension to total security by any military device—whether purported to be a sure deterrent or a leak-proof defense against all attacks—is foolhardy. The competition for military superiority is endless. It has brought civilization to a state of peril never known before. To reduce the competition is more hopeful than to win a short-term superiority in it. The desperate craving for military power—a craving that many nations indulge in—can lead to disaster. In a world where life can never be risk-free, some risks taken for peace are better than the constant risks taken for war.

REFERENCES

1. BOYLE, F. A. 1987. Implementing the Biological Weapons Convention. Genewatch 4(4-5): 12.
2. BLAKESLEE, S. 1989. Panel of scientists fears global race in biological arms. N.Y. Times. January 18: p. A7.

3. Committee for Responsible Genetics. 1988. Pledge against the Military Use of Biological Research. Boston, MA.
4. BROWN, H. 1978. The Human Future Revisited: The World Predicament and Possible Solutions: 59. W. W. Norton. New York, NY.
5. PETERS, T. & S. A. WEISS. 1989. Down with bureaucrats (op-ed). N.Y. Times. January 8: sect. 4; p. 29.
6. BUTTERFIELD, H. 1949. Christianity and History: 104. G. Bell & Sons. London.

The Moral Arguments and the Philosophy of Science and Technology

PAUL T. DURBIN

Department of Philosophy
University of Delaware
Newark, Delaware 19716

Philosophers and social theorists since Karl Marx have worried about technology or the inseparable link between science and technology. At first only a few philosophers considered problems of technology, but since the 1930s, and increasingly in the subsequent decades, problems of technology have come to the forefront of philosophical concern. Many philosophers in the past decade, for example, have contributed to discussions of nuclear deterrence strategy and to discussions of the wisdom of the Strategic Defense Initiative. My own involvement in the philosophy of technology movement (presumably the reason for my being asked to contribute this paper) began in the 1960s; over the past 25 years, my approach has changed very little.

My approach owes the most to the thought of George Herbert Mead.[1-3] According to Mead:

> The order of the universe that we live in *is* the moral order. It has become the moral order by becoming the self-conscious method [of social problem solving] of members of a human society. . . . The world that comes to us from the past possesses and controls us. We possess and control the world that we discover and invent. And this is [to repeat] the world of the moral order.[3]

More usual approaches to moral theory divide the ethical world between deontologists (Immanuel Kant is usually taken to be the best example) and consequentialists (utilitarians are a good example). But if Mead were to agree with Kant that moral behavior transcendentally presupposes the concept of duty, he would at the same time disagree in a profound way with Kant: giving a morality of duty any real meaning requires that our duties have content, and determining the right way to do our duty is not something given in advance—it is something to be worked out in a social dialogue or struggle of competing values. On the other hand, Mead flatly disagrees with utilitarianism as a foundation of morality: there is no way, he says, that the traditional utilitarian concept of self-interest can adequately ground the altruism that is essential to social action.

Descending from these generalities toward the concrete problems addressed here, I will concentrate on the activists who have advocated social responsibility in almost all the scientific and engineering societies since the early 1970s. The way I would construct a Meadian approach to an ethics of science and engineering would be to push scientific and other professionals and their organizations toward greater social responsibility, in the hope of alleviating at least the worst social evils associated with technology. And, as with any group or society, social action runs the risk of stagnation and retrogression unless there is a creative minority that will lead the community toward a better world.

This approach is almost certain to run into problems if it is applied to the research and development communities involved in weapons (and other) research for the military. There are whistle-blowers in the Pentagon, and some of them have become famous for uncovering ethical problems in the weapons procurement system. But are there enough of them, and of other socially responsible activists in the military research community, to exercise the needed ferment function?

More generally, research for or under contract to the military raises questions for me about the ability of conscientious activists to influence policy when research is justified in the name of national security. This can be concretized in terms of Bernard Roth's moral arguments against doing research for the military.[4] And my concern should be obvious: Will such arguments—given the sheer size of the military bureaucracy[5] and the widespread social support for "national defense"—ever get a hearing? (An aside: I even worry whether Robert Dinegar's promilitary religious ethics[6] will get a hearing in the din of arguments based on expediency or claims of overwhelming threats to national security.)

In my advocacy of public-interest activism by engineers and scientists aimed at making their institutions more socially responsible, I have never before faced the challenge of addressing my exhortations to people who might be faced, at every turn, with the challenge that they might be undermining national security. (At the extreme, of course, they have been accused of more—of treason, or at least of "playing into the hands of our enemies.")

One of the main tenets of my approach to ethics is an advocacy of the tolerance and openness associated with a respect for civil liberties. But this too confronts a long tradition, even in the best of administrations, of suspending civil liberties in the name of national security. (Presumably such suspensions are justified as extensions of similar suspensions made in times of war.) If my readings of the defenses of such suspensions of civil liberties are correct, they imply that the Constitution's delegation of the powers of commander in chief to the president supersedes the Bill of Rights. These defenses include assertions that the Constitution allows, at the least, temporary suspensions of some provisions in the Bill of Rights "for the common defense." It might, by further extension, then be said that research in the name of national security enjoys a similar set of privileges. Acceptance of this extension, however, would compromise the right of dissent in research communities working on national security projects.

Perhaps, someone might say, dissenting views could be circulated outside the military. For this to happen, citizens (perhaps some of them military researchers) would have to raise issues of social responsibility, in public debate, with respect to particular weapons development projects. Ordinary citizens, however, seldom know much about weapons development projects until they are nearly completed. Furthermore, the same problem affects public dissent with respect to military developments as impedes dissent within the ranks of military researchers. Public discussions of weapons policies are almost always couched in the language of threats to national security—or even survival. (This language is employed by those on the left as well as by those on the right; that is, doves also worry about the survival of the nation, though they are usually thinking of the survival of a national commitment to the Bill of Rights and other civil liberties.) Is it possible to maintain an atmosphere of civil discussion when each side believes that losing the argument would threaten the defense of a free society (or the survival of the world, including the free world, in the face of a nuclear catastrophe)?

Although some may think it impractical to call for civility in discussions of society-threatening issues, I believe that civilized public debates of these issues are essential. That is why I agreed to contribute a paper to this volume. This volume discusses the moral pros and cons of research for the military, and such discussions are long overdue. Also, this volume includes contributions by military researchers, and such contributions are particularly welcome in debates on these issues.

I also believe, however, that sometimes ethics demands more than polite discussion. If particular groups, and there is no reason that military researchers cannot be among them, feel deeply that specific weapons development projects are unethical or are opposed to the values of a free society, and if normal discussions seem to be getting nowhere in changing the course of development, it may well be necessary to mount public protests, including peaceful and nonviolent demonstrations. If the protestors are particularly incensed and the weapons development policy seems particularly outrageous, I can even imagine another mass political movement of the sort that opposed the war in Vietnam.

In such cases, where people feel that ethics demands public confrontation of military policy, the protestors, if they have any sense, should recognize that defenders of the particular policy will inevitably also come forward. The protestors, as happened during the Vietnam War, may even be accused of treason. And the defenders of military policy are also likely to dress their arguments in the clothing of flag and country—even, on occasion, defending their views with strong ethical arguments.

Polite discussion may be preferable to public protests and counterprotests, but in my view ethics sometimes demands public protest, even national soul-searching, if particular public policies seem egregiously opposed to national values.

In this volume, moral arguments for and against doing research for the military have been debated. My hope is that public debates of such arguments will become a matter of course in this country from now on. In other words, my chief worry is that moral debates may be considered irrelevant in the face of perceived threats to national security. And in such debates, if they take place, I hope that at least some members of the military research community will speak out publicly for what they think social responsibility demands. At times it may be much harder for them than it is for scientists and engineers in industry or in other government institutions; in that case, it is my fervent hope that there will always be activist groups of citizens with whom they can join their voices so that all together they can get their arguments heard.

REFERENCES

1. JOAS, H. 1985. G. H. Mead: A Contemporary Reexamination of His Thought. MIT Press. Cambridge, MA.
2. SLEEPER, R. W. 1986. The Necessity of Pragmatism: John Dewey's Conception of Philosophy. Yale University Press. New Haven, CT.
3. MEAD, G. H. 1964. Scientific method and the moral sciences (1923). In Selected Writings. A. Reck, Ed.: 266. Bobbs-Merrill. Indianapolis, IN.
4. ROTH, B. 1989. The moral arguments against military research. Ann. N.Y. Acad. Sci. This volume.
5. DURBIN, P. T., Ed. 1990. Critical Perspectives on Non-Academic Science and Engineering. Lehigh University Press. Bethlehem, PA.
6. DINEGAR, R. H. 1989. The moral arguments for military research. Ann. N.Y. Acad. Sci. This volume.

PART II. HISTORICAL AND CONCEPTUAL BACKGROUND

Hephaestus and History

Scientists, Engineers, and War in Western Experience

ALEX ROLAND[a]

Military History Institute
United States Army
Carlisle, Pennsylvania 17013

Hephaestus, the god of fire, the armorer of Achilles, was the only ugly god on Olympus. He was also the husband of the most beautiful goddess, Aphrodite. He had powerful arms and lame feet. He produced instruments of war and accoutrements of peace. He was comical and sublime, cast down from Olympus in his youth and restored in his maturity. In him is manifest the Greek penchant for contradiction and irony.[1] In him too is manifest an ambiguity toward the manufacture of instruments of warfare that has troubled Western civilization for the last three millennia. Is it ugly or sublime, Hephaestus still asks, to forge weapons?

For most of Western history the question was mute if not moot. Only in the nineteenth and twentieth centuries have Westerners opened a public debate on the Hephaestus question. The long silence has been occasioned by two factors, one specific and peculiar to science and engineering, the other general and dictated by the cultural evolution of the West. Only when scientific and engineering communities appeared could there be a proper forum for such a debate. Only when Western society grew truly alarmed about the technology of war could the debate become widespread and urgent. For reasons I hope to lay out below, these conditions did not begin to obtain until the late nineteenth century; they did not become ripe until 1945.

The timing of the debate is of crucial importance. Had it taken place earlier, the agenda might have been very much different. Indeed, the discussion now is hardly a debate at all. Rather those who would impose constraints on the service of science and engineering in war preach to each other and to the heathen, looking to stoke the choir or convert the innocent. Those who would defend or even expand the service of science and engineering in war keep their own counsel or speak in muted tones among themselves. Their argument is not necessarily weaker or less sincere. Rather it struggles against what Carl Becker called a "climate of opinion . . . , those instinctively held preconceptions in the broad sense, that Weltanshauung or world pattern," that

---

[a] Present address: Department of History, Duke University, Durham, North Carolina 27706.

can become so deeply imbedded in a cultural consensus that it is not even spoken any more or recognized consciously.[2]

In this paper I hope to trace the evolution of these two phenomena—the emergence of professional scientific and engineering communities and the development of the modern consensus on technology in war—and show how their timing contributed to the current climate of opinion and the debate it has spawned. I will argue that the timing of the debate virtually predetermined the nature of the discussion.

. . . .

Who were the scientists and engineers who might have engaged in the development of weapons? For the engineers, at least, the answer is practically all. The first engineers, so called, were military engineers, the men who devised, built, and operated engines of war. These originated with the Sumerians and Assyrians and came to their earliest perfection with the catapults and ballistae of the ancient Greeks. The term engineer in its modern form was first applied in the Middle Ages, derived from the Latin root for ingenious and ascribed to those whose natural talents allowed them to create ingenious machines of war. Engineers did not take their name from engines; both the machines and their makers took their names from being ingenious. Those who engaged in other forms of what we would now call engineering were most often referred to as architects, the term extolled by Vitruvius in his first-century B.C. classic *De Architectura*.[3] The clear distinction between civil and military engineering did not occur until the middle of the eighteenth century, when John Smeaton first described himself as a civil engineer.[4] Even then, however, professionalization was still more than a century away. Not until the late nineteenth century did engineering societies begin to emerge and establish professional standards.[5]

Scientists followed a similar path. If we understand scientists to be those who seek an understanding of the physical world as an end in itself, then surely there were such people in prehistoric times. Indeed the *Dictionary of Scientific Biography* has entries for individuals as far back as ancient Babylon.[6] But the term scientist was not coined until 1840,[7] and even then there was no professionalization in the sense we would understand today. Surely the scientists had formal institutions long before the engineers; the Royal Society of London and its precursors date from the seventeenth century.[8] But true professionalization, as with the engineers, began only in the second half of the nineteenth century.[9]

Before these terms took on their modern meaning, there were many instances of what we would now call scientists and engineers working for purposes of war. The ancient engineers whose handiwork is described by J. G. Landels certainly displayed a remarkable degree of sophistication even before the birth of Christ.[10] Dionysius of Syracuse is reported to have set up what we would now call a government arsenal, where he engaged what we would now call scientists and engineers to conduct what we would now call research and development in weaponry.[11] Archimedes, as best we know, did not scruple to use his talents for purposes of war, though there is an argument that he was worried about dirtying his hands in the development of practical machines, which were considered the province of slaves.[12] Vitruvius and Frontinus seemed as proud of their civil engineering as they were of their military engineering.[3,13] Kallinikos, the legendary inventor of Greek fire, a kind of early napalm, was supposedly an architect who willingly brought his invention to Constantinople so that it might be used to save Byzantium from the Moslems.[14]

Not until the time of Leonardo do we find a well-documented case of a scientist or engineer who scrupled to engage in military research and development. Even there, however, Leonardo was concerned only about his design for a submarine, which he would "not publish or divulge on account of the evil nature of men." Otherwise he was more than satisfied to style himself a military engineer and to use his talents for purposes of war, if that was what was necessary to secure himself patronage.[15] Galileo could not in good conscience bend his scientific observations to the teachings of the Church, but his conscience was untroubled by work at the arsenal in Venice and by other contractual work for the military.[16] The scientific revolution of the eighteenth and nineteenth centuries witnessed an explosion of scientific activity and a comparable increase in the commitment of scientific resources to purposes of war. The wars of the French Revolution and Napoleon became justly famous for the enlistment of distinguished scientists like Gaspar Monge; Pierre Simon, Marquis de Laplace; Jean-Baptiste-Marie Meusnier; and G.-Riche de Prony—who were all part of Carnot's plan to bend the total resources of the state to the war effort. Many of these savants came to the service of the war effort through their earlier connections with the schools of military engineering pioneered by the French in the waning decades of the Old Regime.[17] These too became models for other states in the nineteenth century.

Thus, before scientists and engineers emerged as modern professionals in the late nineteenth century, they already had an unbroken record of service in war dating all the way to Hephaestus and beyond. It was not until professionalization, however, that scientists and engineers had the forum and the impetus to begin discussing among themselves and with society at large their moral and ethical responsibilities as a profession. Thus it was that the Hephaestus question came to be debated in an atmosphere that drove the discussion all in one direction.

• • • •

How that atmosphere or climate of opinion formed is a complex issue of Western historical experience. Ask a dozen historians and you are likely to get a dozen different candidates for the most important historical determinants. I have chosen to address five of them, not because I am sure that these explain all, but because I believe that they can, in short compass, suggest the kinds of forces at work in shaping our current debate. This list, therefore, is meant to be suggestive, rather than comprehensive and definitive.

Chivalry and the international law of war that flowed from it established our modern notions of the *jus in bello*, that is, the means by which it is appropriate to conduct war.[18] Like all restraints on warfare, the code of chivalry was designed in the first instance to protect the in-group, in this case, Christian knights.[19] It did not protect the infidel; the famous papal injunction against the use of the crossbow did not apply to non-Christians. Nor did the code of chivalry necessarily protect Christian civilians. The Church tried first a Peace of God and then the less successful Truce of God in an attempt to keep the private wars of the medieval knights from spilling over into the civilian community, where tithes were paid.[20] Failing that, the Church helped promote the Crusades, which would at least send the knights off to fight Muslims instead of leaving them in Europe to kill each other and chew up the tax base.

Religion was the glue that held this culture together, and it was as successful as any institution has ever been in controlling warfare. The Church usually addressed the *jus ad bello*, that is, the laws surrounding the initiation and termination of war. Less often did it speak to the *jus in bello*, except to sustain the medieval notion of a judicial duel. Warfare between Christian knights was viewed as a contest to determine the will of God. All things being equal, the anointed of God would emerge triumphant. As the sixteenth century jurist Alberico Gentili stated it, "the 'magic arts' are . . . unlawful in war, because war, a contest between men, through these arts is made a struggle of demons."[21] Ersatz tactics or exceptional weapons, of the kind that might be provided by scientists, engineers, or other malefactors of the devil, were viewed as mischievous innovations that might distort the manifestation of God's will. The crossbow was banned because it allowed a mere commoner the power to bring down a knight; surely that could not be God's will.

International law, as it emerged in the early modern period, functioned in much the same way. It was originally based in the Church, and it took Christianity to be the common bond of the participating nations. Those outside the Christian community were not expected to abide by the rules, nor were they entitled to the restraints that Christians employed with one another. As in medieval chivalry, the prince could always suspend the rules of war for what were called necessities of state, meaning that the rules applied unless the state faced disaster—then it was no holds barred.[22] International law, especially in its early years, was silent on the means of war, except to the extent that it inherited chivalric sanctions against such actions as surreptitious murder and other forms of treachery. These last restraints were again class based, an attempt to ensure that the warrior class would fight by known methods that would limit the danger.

The transition from medieval restraints on war to a nascent international law was shaped in part by humanism, another of the factors contributing to our modern notions of the proper role of scientists and engineers in war. Tied closely to the Reformation of the sixteenth century, the humanist movement eschewed the fatalism and determinism of medieval Christianity to argue instead for the power and importance of the individual and the sanctity of human life. Just as Luther proclaimed the ability and the right of individuals to read and interpret the Bible for themselves, so too did the humanists argue for the ability and the right of individuals to find peace and happiness on earth, no matter what heaven they might be headed to. Life on earth did not have to be, in fact should not be, a vale of tears endured on the way to some promised reward. War was not necessarily an inevitable part of the human condition, and even when war did take place, it was no license for the slaughter of innocents. All humans had a right to life and concomitantly a moral responsibility to respect and protect the lives of others. Here, even more than in international law, began the notion of moral responsibility to the human community at large that now moves many scientists and engineers to concern for the military consequences of their work.

In the seventeenth and eighteenth centuries, these forces of chivalry, international law, and humanism were compounded by an emerging despair over the declining efficacy of war. However cruel and disastrous warfare might have been in human history, it usually had the redeeming virtue of settling human disputes. War was, as Clausewitz described it, a continuation of politics by other means.[23] Nations that could not solve their disputes amicably contended in arms and accepted the outcome for better or worse as the normal course of events. Medieval knights even saw the outcome as God's will, to which victor and vanquished alike resigned themselves. Battles were limited in time and space and they tended to settle the issue. The Wars of Religion, however, changed that perception dramatically.[24] They were unprecedented in European experience for longevity, cruelty, civilian casualties, and indecisiveness.

The revulsion generated by the Wars of Religion has become a more or less constant ingredient of Western civilization, cast in the shadows from time to time by a Napoleon or a Moltke or a John Wayne, but never far from the center. It was reflected in the eighteenth century by the limited and circumscribed warfare of dynasty and empire and by the biting commentary of writers like Voltaire, whose *Candide* remains one of the great antiwar tracts of Western culture. The fact that George Will can now call Ronald Reagan the Pangloss of modern international relations suggests the hold that this intellectual current still has on our cultural outlook.[25] Though submerged for much of the nineteenth century in the illusion created by Napoleon and the Prussians who imitated and institutionalized him, the perception reemerged undiminished in the twentieth century in the writings of such diverse authors as Ivan Bloch,[26] Norman Angell,[27] Walter Millis,[28] Quincy Wright,[29] and John Keegan.[30]

Despair over the decreasing efficacy of modern war was accelerated in the nineteenth century by the increasing deadliness of war, a trend that has produced its own literature.[31] Three nineteenth-century revolutions brought on this increasing lethality of war.[28,32] The democratic revolution, originated by the Americans but consummated by the French, unleashed the nation in arms, an innovation unthinkable in the eighteenth century, when an armed populace might as likely turn on its own government or its military leaders as on the enemy. Only when the masses came to believe, however erroneously, that the government and its causes were theirs was it possible to place modern firearms in their hands and expect them to use them for purposes of state. An industrial revolution was needed to supply the *levee en mass* made possible by this political revolution. And at the critical moment one came along, first in England, then in the United States, Germany, and France—the very countries that would inaugurate the total war to which these successive revolutions were leading. First, however, it was necessary to invent yet a third revolution to develop the means to administer the huge and increasingly mechanized armies that were taking the field in the nineteenth century. For this the Prussians developed a general staff system, a managerial revolution that was so successful in the wars of German unification that all the major Western powers instituted similar organizational reforms by the time World War I broke out. Modern total war was born of these three nineteenth-century revolutions.

Total war, as it was practiced in the first half of the twentieth century, in turn created the impression that the technology was principally responsible for the unimaginably increased deadliness and pointlessness of war, perceptions that directly fed the emerging sense within the scientific and technical communities that this process must be arrested. In an age of intense nationalism, the mass armies spawned by the democratic revolution were simply taken for granted. The managerial revolution was ignored. The technological revolution was left as the most obvious cause of the horrors of total war. In fact, the world wars were really wars of industrial mobilization, in which valor, training, tactics, and strategy had less to do with the outcome than the industrial strength of the competitors. The United States was the decisive combatant in both wars because it had the capacity to simply drown the enemy in resources, to destroy the enemy's will and means to resist by engaging in a huge and enormously destructive attrition of industrial power.

But even as industrial capacity was determining the outcome of the world wars, the perception was arising that quality and not quantity would determine the outcome of the next war. World War I was in some senses the chemists' war, fought in the laboratories that were developing poison gas and high explosives.[33] World War II was the physicists' war, shaped by radar, sonar, aerodynamics, the internal combustion engine, and communications.[34] In practice the chemists and physicists worked hand-in-glove with engineers, to the point where their contributions were indistinguishable. The scientists increasingly needed engineers both to provide the equipment with which they conducted their empirical research and to provide the know-how for reducing their ideas to practical applications. Comparably, the engineers needed the scientists both to provide conceptual ideas of what might be done and to troubleshoot the execution of what was being done. It was perfectly appropriate that the World War II Office of Scientific Research and Development should be headed by an engineer, Vannevar Bush, and that the head of the wartime laboratory at his home institution, the Massachusetts Institute of Technology, should be a scientist, Lee DuBridge.[35]

The trend in scientific and engineering contributions to warfare climaxed in the Manhattan Project, which drew together scientists and engineers on a military undertaking of unprecedented scope and consequences. The successful development of the atomic bomb elevated science and engineering to unheard-of influence and prestige. It reinforced the notion, already rampant in military circles, that future wars would be decided by the quality of science and engineering rather than the scale of industrial production. It dragged scientists and engineers into a more or less permanent state of preparation for war. And it raised the prospect that the next round in the escalating application of science and technology to war might well exterminate human life. Because it raised so poignantly the trends that had been emerging in the West since the late middle ages—chivalry, international law, humanism, the declining efficacy and increasing deadliness of war—the atomic bomb has become a symbol for our age and a stimulus to the debate that now rages over the proper role of scientists and engineers in war or preparation for war.

The difference in the last forty years has been that the scientific and engineering communities now have institutional arrangements both for defining themselves professionally and for determining the appropriateness of military work. The *Bulletin of the Atomic Scientists* is representative of the institutional apparatus that is now available not only to conduct a debate on the proper role of scientists and engineers in war but in fact to stimulate the debate. Some now argue that scientists and engineers should not lend themselves to this kind of work—surely not in peacetime, probably not in war. Others argue that scientists and engineers should. The perception that science and technology will be the arbiters of the next war does not silence the debate; it only heightens it. None of the arguments on either side are new.

Those opposed to the participation of scientists and engineers in military research and development tend to cite the Western cultural developments that have been delineated in this essay. They protest against the development of what are now called "dubious weapons," that is, those that seem to give one side or another an unfair or immoral advantage.[36] They cite the moral obligation of all humans to prize the sanctity of human life and to accept their responsibility, as members of the human community, to eschew action that cheapens or threatens human life. They note the increasing deadliness of warfare and its decreasing efficacy, especially in the modern age of total war and, finally, nuclear war. They look upon the new concoctions of science and technology, such as Star Wars, stealth aircraft, and smart bombs, as still greater abominations. Always they look back at Hiroshima as the inauguration of weapons of mass destruction, a watershed of sorts in the application of science and technology to purposes of war, a final warning perhaps of where this line of development would lead.

Scientists and engineers on the other side of the aisle make arguments with equally long historical pedigrees. They suggest that we must make war so terrible that humans will finally abandon it, a goal and a rationalization that goes back at least as far as the ancient Greeks. John Donne, whom no one would mistake for a warmonger, extolled the virtues of cannons in 1681, claiming that since their introduction wars were shorter and hence less deadly and destructive.[37] Advocates also cite the need to bend their efforts to just wars, lest the forces of evil triumph in the world. The atomic bomb was developed in the first place, after all, out of fear that Hitler, the ultimate devil of modern times, might secure one for himself. Still others argue with cavalier detachment that science is neutral and that it is not encumbent upon scientists to concern themselves with the eschatology of their discoveries. As Richard Feynman, the Nobel Laureate physicist, put it when asked if his work did not make him feel good because it was for peace: "No, that never enters my head, whether it is for peace or otherwise. We don't know."[38] Finally, it is argued that spin-offs from military research and development profit civilian society,[39] a case that is being made with greater and greater frequency in the 1980s as a larger and larger portion of the research and development in the United States is funded by the Department of Defense.[40] It was essentially the argument of Werhner von Braun, who claimed after the fact that his development of the V-1 and V-2 weapons in World War II was really an exploitation of the Nazis in order to advance the cause of spaceflight development.[41]

For the present, at least, those who would circumscribe the role of scientists and engineers in war strike the more resonant chord in our culture. They are in tune with the climate of opinion that reigns in the world today. That climate has been building at least since the days of chivalry and it will not soon be dispelled. In fact, much of the last 500 years of Western experience can be seen as confirming the central wisdom of that world view.

Confirmation of the opposite view awaits some future development. Current philosophical arguments about good scientists and good engineers triumphing over evil counterparts, or neutral scientists and neutral engineers in the service of virtuous states, had some currency at midcentury in the age of the totalitarian state, but they are losing their power in a world of détente and the beatification of the evil empire. Looking to the future, we may already be in a transitional age when fear of nuclear war will force nations to find nonviolent ways to resolve their differences. Should that trend continue, scientists and engineers who built the weapons of the last fifty years may be seen by future generations as peacemakers. Similarly, Star Wars or some such technology may one day establish the superiority of defense over offense and thus bring security to all. Computers may turn out to be a democratizing technology that empowers individuals and renders states vulnerable. But in the absence of these developments, the weight of Western experience will continue to give pause to Hephaestus and his heirs.

REFERENCES

1. DWIGHT, M. A. 1872. Grecian and Roman Mythology: 158–165. A. S. Barnes. New York, NY.
2. BECKER, C. 1932. The Heavenly City of the Eighteenth-Century Philosophes: 5. Yale University Press. New Haven, CT.
3. VITRUVIUS, M. 1860. The Architecture of Marcus Vitruvius Pollio. J. Gwilt, Trans. J. Weale. London.
4. PACEY, A. 1975. The Maze of Ingenuity: Ideas and Idealism in the Development of Technology: 204–212. Holmes and Meier. New York, NY.
5. LAYTON, E. T., JR. 1986. The Revolt of the Engineers: Social Responsibility and the American Engineering Profession. Johns Hopkins University Press. Baltimore, MD.
6. GILLISPIE, C. C., Ed. 1970–1980. Dictionary of Scientific Biography (16 vols.). Scribner. New York, NY.
7. Oxford English Dictionary. 1971. *See* scientist.
8. ORNSTEIN, M. 1928. The Role of Scientific Societies in the Seventeenth Century. University of Chicago Press. Chicago, IL; MCCLELLAN, J. E., III. 1985. Science Reorganized: Scientific Societies in the Eighteenth Century. Columbia University Press. New York, NY.
9. BEN-DAVID, J. 1971. The Scientists's Role in Society: A Comparative Study. Prentice-Hall. Englewood Cliffs, NJ.
10. LANDELS, J. G. 1981. Engineering in the Ancient World. University of California Press. Berkeley, CA.
11. FERRILL, A. 1985. The Origins of War: From the Stone Age to Alexander the Great: 170–171. Thames & Hudson. London.
12. BERNAL, J. D. 1965. Science in History. 3rd edit. Hawthorn Books. New York, NY.
13. FRONTINUS, S. J. 1969. The Strategems and the Aqueducts of Rome (original 1925). C. E. Bennett, Trans. M. B. McElwain, Ed. William Heinemann. London.
14. PARTINGTON, J. R. 1960. A History of Greek Fire and Gunpowder: 12. W. Heffer & Sons. London.
15. DA VINCI, L. 1962. The notebooks of Leonardo da Vinci (2 vols.). E. MacCurdy, Trans. Vol. 2: 109. Reynal & Hitchcock. New York, NY.
16. DRAKE, S. 1978. Galileo at Work: His Scientific Biography. University of Chicago Press. Chicago, IL.
17. GILLISPIE, C. C. 1980. Science and Polity in France at the End of the Old Regime. Princeton University Press. Princeton, NJ.
18. KEEN, M. H. 1965. The Laws of War in the Late Middle Ages. Routledge & Kegan Paul. London.
19. DAVIE, M. 1968. The Evolution of War: A Study of Its Role in Early Societies (original 1929). Kennikat Press. New York, NY; TURNEY HIGH, H. H. 1971. Primitive War: Its Practice and Concepts. 2nd edit. University of South Carolina Press. Columbia, SC.
20. BALLIS, W. 1973. The Legal Position of War: Changes in Its Practice and Theory from Plato to Vattel. Garland. New York, NY.
21. GENTILI, A. 1933. De Iuri Belli Lubri Tres (original 1612). J. C. Rolfe, Trans: 160–161. Clarendon. Oxford.
22. BEST, G. 1980. Humanity in Warfare: The Modern History of the International Law of Armed Conflict. Weidenfeld & Nicolson. London.
23. VON CLAUSEWITZ, C. 1976. On War. M. Howard & P. Paret, Eds. Princeton University Press. Princeton, NJ.

24. NEF, J. U. 1968. War and Human Progress: An Essay on the Rise of Industrial Civilization (original 1950). Norton. New York, NY.
25. WILL, G. 1989. How Reagan Changed America. Newsweek. January 9: 15.
26. BLOCH, I. S. 1900. Modern Weapons and Modern War (an abridgment of The War of the Future in Its Technical, Economic and Political Relations). G. Richards. London.
27. ANGELL, R. N. 1910. The Great Illusion: A Study of the Relation of Military Power to National Advantage. William Heinemann. London.
28. MILLIS, W. 1956. Arms and Men: A Study in American Military History. Putnam. New York, NY.
29. WRIGHT, Q. 1965. A Study of War. 2nd edit. University of Chicago Press. Chicago, IL.
30. KEEGAN, J. 1976. The Face of Battle. Viking Press. New York, NY.
31. BODART, G. 1916. Losses of Life in Modern Wars: Austria-Hungary; France. Clarendon Press. Oxford; RICHARDSON, L. F. 1960. Statistics of Deadly Quarrels. Q. Wright & C. C. Lienau, Eds. Boxwood Press. Pittsburgh, PA.
32. FULLER, J. F. C. 1961. The Conduct of War, 1789-1961: A Study of the Impact of the French, Industrial, and Russian Revolutions on War and Its Conduct. Rutgers University Press. New Brunswick, NJ.
33. HARTCUP, G. 1988. The War of Invention: Scientific Developments, 1914-1918. Brassey's. London.
34. KEVLES, D. J. 1978. The Physicists: The History of a Scientific Community in Modern America. Knopf. New York, NY; VAN CREVELD, M. 1989. Technology and War: From 2000 B.C. to the Present. Free Press. New York, NY.
35. BURCHARD, J. 1948. Q.E.D.: MIT in World War II. Wiley. New York, NY.
36. Stockholm International Peace Research Institute. 1976. The Law of War and Dubious Weapons. Almqvist & Wiksell. Stockholm.
37. DONNE, J. 1839. Sermon CXVII. *In* The Works of John Donne, D.D., Dean of St. Paul's, 1621-1631, with a Memoir of His Life (6 vols.). H. Alford, Ed. Vol. 5: 58. J. W. Parker. London.
38. FEYNMAN, R. 1986. What Do *You* Care What Other People Think? More Adventures of a Curious Character: 86. Norton. New York, NY.
39. SOMBART, W. 1975. Krieg und Kapitalismus (original 1913). Arno Press. New York, NY.
40. NOZETTE, S. 1986. The Commercial Potential of SDI. *In* Promise or Peril: The Strategic Defense Initiative. Z. Brzezinski, Ed.: 189-197. Ethics and Public Policy Center. Washington, DC.
41. ORDWAY, F. I., III & M. R. SHARPE. 1979. The Rocket Team. Crowell. New York, NY.

Drawing the Line
An Examination of Conscientious Objection in Science

ROSEMARY CHALK[a]

Institute of Medicine
National Academy of Sciences
Washington, DC 20016

INTRODUCTION

Shortly after the end of World War II, in 1946, the American Quaker pacifist A. J. Muste initiated a series of letters to Albert Einstein. Muste's first letter, which he circulated broadly, was in response to a public appeal by Einstein's Emergency Committee of Atomic Scientists urging the public to donate funds to support the scientists' efforts to develop national and international controls limiting the use and development of nuclear weapons. Throughout his correspondence with Einstein, Muste argued that if the scientists' message was to be taken seriously by the general public, it needed to be accompanied by an expression of moral conviction by those deeply concerned with the problems presented by atomic weapons. He called upon Einstein and other atomic scientists to renounce publicly any further involvement in building these weapons:

> ... the scientists are continuing to make atomic and biological weapons. Even if they do not participate in the actual current production, they serve on commissions which plan and supervise production. And even those leading scientists who are not thus engaged are still implicated so long as they do not openly say to the world and especially to their scientific colleagues that it is unworthy of a scientist and an honest man to have any part in the insane and treasonable business of taking part in an armaments race which provides no defense ... and which will lead straight away from peace and into the nightmare of atomic war.[1]

Muste concluded the letter as follows:

> As for the masses, how can they be expected to believe that atomic weapons are as worthless and horrible as the scientists say they are, when the scientists continue to make the things? It cannot make sense to ordinary human beings.[1]

---

[a] Institutional affiliation is listed for purposes of identification only and does not imply that the views expressed in this paper are those of the Academy. The author wishes to acknowledge the support and assistance of the Science, Technology, and Society Program at the Massachusetts Institute of Technology in preparing research material for this paper. Most of the research was conducted during the period 1982-1983 when the author was in residence at the Institute's Science, Technology, and Society Program as an Exxon Research Fellow.

Muste's call was consistent with his own commitment to pacifism and conscientious objection. He had previously formed two organizations founded on these beliefs, the Fellowship of Reconciliation and the War Resisters League, and he had consistently argued against personal participation in any preparation for war. He was the leader of a small group of Quakers who advocated conscientious objection as the only appropriate response to military conscription, even after the bombing of Pearl Harbor.[2]

Muste's open letter to Einstein and other correspondents generated a number of responses from the atomic scientists involved with Einstein's Emergency Committee. Some of the replies were supportive, such as Einstein's own statement, that "your criticism seems to me justified to a high degree,"[3] and a reply from Robert Hutchins (the Chancellor of the University of Chicago), who wrote that "a movement is gaining ground among scientists . . . against working on anything that looks like a weapon."[4]

Others were more skeptical. Leo Szilard wrote that he agreed with the moral issue raised by Muste, but noted that "considerations of expediency" on the part of "the majority of scientists" would doom Muste's proposal to failure.[5]

Some of the atomic scientists responded to Muste's letters by stating that it was not the responsibility of scientists to determine on their own whether such research should continue. For example, Harold Urey stated that "I personally believe in obeying the laws of the country. . . . I exercise my right as a citizen in attempting to change the rules, not in frustrating the rules as they are laid down."[6] Hans Bethe indicated that a "strike" by the scientists "would only antagonize the public of the United States who would rightly accuse us of trying to dictate the policies of the country."[7] Willie Higinbotham, the chairman of the newly organized Federation of American Scientists, was even more emphatic: "We believe in government by the people . . . if scientists were to walk out on all military projects they would be taking the law into their own hands just as surely as the Ku Klux Klan."[8]

This exchange of letters between Muste and the atomic scientists raises several issues that deserve to be addressed in considerations of ethical issues associated with military research. The idea of drawing the line between scientific work and military objectives did not originate with Muste, but his correspondence illustrates several philosophical and political questions regarding the nature of scientific responsibility with regard to military research.

In examining these questions, I will focus on the issue of conscientious objection in science: I will describe some examples of extreme and modified forms of conscientious objection in science, illustrate responses to these positions, and identify several issues that deserve further analysis.

In presenting this material, I am not recommending conscientious objection as the only or even the best form of expressing moral concerns about the relationships between science and the military. But these actions constitute a phenomenon that has appeared with some regularity throughout the history of science, a phenomenon that has not been fully developed in the literature on science and social responsibility. I suggest that an analysis of conscientious objection among scientists will identify many difficult points about the broader limits and relationships between individual and social responsibility in science.

ORIGINS OF CONSCIENTIOUS OBJECTION

Conscientious objection is most commonly rooted in a particular religious or moral philosophy, such as that of the Quakers or the Mennonites.[9] More recently, conscientious objector status has been sought by individuals objecting to the draft on the basis of individual conscience alone, without reference to a formally organized religious tradition. In past centuries, conscientious objectors often experienced severe penalties for their beliefs, in the form of imprisonment, fines, or social ostracism.

In the twentieth century, conscientious objectors have been granted forms of alternative service during all major U.S. wars, although grudging tolerance probably best describes their position in American history. I am not certain, but I believe that the law does not explicitly recognize the rights of conscientious objectors on a universal basis, but leaves the treatment of such cases to the discretionary judgment of local draft boards.

Explicit legal protections for conscientious objectors were considered early in American history, however. One study suggests that an early draft of the Constitutional Bill of Rights prepared by James Madison included the right to refuse to bear arms in addition to the right to possess them.[10] This proposed section was discarded in the series of political compromises that accompanied the final ratification of the first amendments of the U.S. Constitution, but a majority of the original states recognized the right to refuse compulsory military service provided refusals were based on religious beliefs.

Conscientious objection, although not called such, has appeared with some consistency throughout the history of science. But the idea of drawing lines in some absolute fashion on military research—that is, research sponsored directly by the military or research with specific military applications— has received little enthusiasm or collective support within the scientific and technical community. There are a few public figures, some of them leading scientists in their fields, who have openly supported the idea that as a matter of general principle scientists should not be involved in any kind of military research. Some scientists modify this position, proposing that scientists should not participate in the development of particularly destructive weapons of war, most commonly with reference to nuclear, biological, or chemical weapons.

SCIENTISTS AS CONSCIENTIOUS OBJECTORS

Three scientists who have publicly renounced military research are the Soviet physicist Peter Kapitza, a Nobel Laureate who spent several years under house arrest for his refusal to work on atomic bomb research under

Stalin[11]; MIT mathematician Norbert Wiener, founder of the field of cybernetics research, who in the aftermath of Hiroshima refused to communicate his earlier military work[12]; and mechanical engineer Victor Paschkis at Columbia University, a friend and associate of A. J. Muste.[13] Paschkis was instrumental in organizing the Society for Social Responsibility in Science (SSRS) and its counterpart, the Society for Social Responsibility in Engineering (SSRE).

Albert Einstein is another major scientific figure who repudiated the military, and he generally urged scientists to cease from engaging in military research.[14] Einstein's position is complicated somewhat by his belief that the conditions presented by Hitler's aggression in Europe required armed resistance by other nations, and the cooperation of scientists in organizing such resistance. His role in initiating the U.S. atomic weapons project is well known, although he later referred to this action as the "great mistake of my life." After Hitler's defeat, Einstein reaffirmed his original pacifism and resistance to military activities, as his reply to Muste, noted above, would indicate.

There have also been public declarations by groups of scientists that include formal statements of conscientious objection. For example, in April 1957, eighteen German atomic scientists published a declaration calling upon the Federal Republic of Germany to renounce the possession of atomic weapons of any kind, and stating, further, that the signatories would themselves not participate in the production, test, or application of any atomic weapon.[15] The signatories included Max Born, Otto Hahn and Fritz Strassman (the discoverers of fission), Werner Heisenberg, and Max von Laue. Born, Hahn, Heisenberg, and von Laue were all Nobel Prize winners.

A statement written by theoretical physicist Carl von Weizsacker later described the events that led to the German declaration:

> We cannot close our eyes to the danger . . . of relying for protection against the Soviet power on a weapon as deadly as the atomic bomb. Public opinion must now assimilate facts which have been long familiar to scientists. However, men can be only imperfectly influenced by words alone. It is not sufficient to tell them that this toy is dangerous. Only if they see that we are ready to make real sacrifices to avoid these dangers will there be a hope that they will be impressed, even in their subconscious minds. This is why it seemed to me important to declare publicly, for ourselves – and if possible also for our country – that we renounce the making and using of atomic weapons.[15]

The SSRS and the SSRE, mentioned earlier, required their members, upon joining the organizations, to sign a pledge stating that they would not engage in research for destructive purposes. Their history and membership are difficult to trace, but the organizations probably numbered about 700 members during the 1950s.[16] The SSRS became particularly active in supporting efforts during the Vietnam War directed toward studying the effects of the use of herbicides in that country, with the objective of limiting or eliminating such use. The SSRS and the SSRE published a newsletter and occasional journal (titled *Spark*), and several of the founding members (including

Paschkis) were active in forming the Committee on Social Implications of Technology (later upgraded to section status) within the Institute of Electrical and Electronics Engineers (IEEE). This IEEE section has subsequently formed several regional chapters, and now publishes a journal titled *Technology and Society*. The concerns of the IEEE group, although stimulated by the pacifist sentiments advocated by many of the early SSRS members, are now focused more broadly on general science and society areas, including privacy issues related to computer use, energy conservation problems, and codes of ethics for engineering societies.

Another example of collective conscientious objection activity is a pledge that was circulated on the MIT campus in the early 1970s during the controversy over whether the Institute should continue to be involved in classified military research. The pledge, drafted by the group known as Scientists and Engineers for Social and Political Action (SESPA), stated:

> I will not participate in war research or weapons production. I further pledge to counsel my students and colleagues to do the same.[17]

THE MEDIEVAL AGE

The idea of conscientious objection in science, though illustrated in the previous examples by references to scientists in the modern postatomic age, is probably as old as the methodology of science itself. Although extensively involved in constructing and designing military weapons, Leonardo da Vinci is said to have deliberately destroyed his designs for an undersea vessel, a potential prototype of the modern submarine, in the belief that such information would only be used for the purpose of human destruction.[18]

Another incident is attributed to British mathematician John Napier, the founder of the theory of logarithms, who had experimented with a new form of artillery design around 1600. This invention, according to the reports of witnesses, cleared a field four miles in circumference, destroying "all living creatures exceeding a foot in height" in the process. Napier took great pains to conceal the workings of his invention, however, and even spoke from his deathbed against devising new weapons:

> For the ruin and overthrow of man there were too many devices already framed which if he could make to be fewer, he would with all his might endeavor to do . . . therefore by any new conceit of his the number of them should never be increased.[19]

A more indirect approach to restricting military research is credited to Robert Boyle, the British physicist who formulated the early laws of the behavior of gases under changes in pressure. In 1680, a correspondent of Boyle's chided him for having invested personal funds in supporting the work of a Dr. Kuffler, a German scientist known for his development of explosives. It seems that some twenty years earlier, a mutual acquaintance of Boyle's and

Kuffler's had quietly told Boyle that the German, although much involved in military inventions, was actually more interested in developing improved methods of cultivating farmland, but could not find a patron to support this research:

> If (Kuffler) could get any partners, or any other encouragement or assistant for it, he would willingly desist from all eager pursuits about his dreadful and destroying invention.[20]

Boyle proceeded to act upon the friend's advice by providing funds for the German's less destructive research interests.

A more recent example of an action similar to Boyle's, but even more indirect, is a statement by the biologist Albert Szent-Gyorgyi, who, after being awarded the Nobel Prize in 1937, sought to invest his award funds in peaceful pursuits in order to avoid corrupting his own scientific interests:

> I went to a broker and asked him to buy shares for me, but told him to buy only shares which in case of war would go down and not up. Knowing the make-up of the human brain, I was afraid that should I have shares which would go up in case of war, I would unconsciously wish for war, and accept as truth any argument which promoted war. I did not want to fall into this trap having worked all my life for peace and the welfare of man. War came, all the same, and I lost my money, but saved my conscience.[21]

One should not conclude from these examples, however, that the idea of conscientious objection in science has had broad appeal or support within the general scientific community. An uncertain number of scientists may adopt the conscientious objector position or some modified form of it, but they often do so privately without stating publicly the reasons for their choice. Many if not most scientists and engineers, however, have been generally skeptical of the merits of this approach. Indeed, scholarly discussions of social responsibility in science categorize Norbert Wiener's position as an extreme and not very popular response to ethical concerns involving military research.

Some individuals have publicly challenged the conscientious objector approach. Lewis Strauss, for example, the first chairman of the Atomic Energy Commission, argued that atomic weapons were essential to the preservation of the basic principles and traditions of American society:

> Is the use of force wrong when it is in defense of freedom and justice? Is the defense of the weak against the strong wrong? If the answer to these questions be "No," then there is no consistent argument against weapons designed to deter assaults against freedom and justice, or aggression against weak populations by brutal governments, simply because they are weapons with force of a new magnitude. Or if justice and the oppressed are to be defended, but cannot be successfully defended with the sword, is the cannon a permissible weapon? And where, after that, is the line—and who may draw it?[22]

Others have dramatically urged scientists to contribute their talents to military research, as suggested in the book *Scientists against Time*:

> The physicist or chemist engaged in war research knows that he is matching his wits against physicists or chemists in enemy laboratories, equally bent on giving to their national forces that measure of technological superiority on which success in modern war largely depends.[23]

And, in an interview during the controversy over the decision to build the hydrogen bomb, Harold Agnew, former director of the Los Alamos Laboratory, made the following statement:

> There were remnants of this feeling that we shouldn't pursue these [atomic weapons]. . . . I was not in sympathy with those individuals; in fact, I thought they were nuts. I just didn't understand them . . . they even got into religious matters. They would quote the Bible. I thought they were quite off their rocker, frankly. . . . My feeling was they ought to leave the place [Los Alamos] because we had a responsibility to the country to explore what could be done.[24]

INDIVIDUAL CHOICES

It should not be assumed that these two extremes adequately express the problem of drawing lines, that is, making decisions about the moral acceptability of military research. Some scientists and engineers, being skeptical of both the patriotic and the conscientious objector approaches suggested in the examples above, address the moral questions of military research more selectively. It is acceptable, they might conclude, to work on conventional forms of defense research that protect the United States from unprovoked aggression. Yet research on offensive projects or weapons targeted against civilians may be considered wrongful.

Such views are expressed in the following letter from an electrical engineer published in the IEEE journal *Spectrum*:

> My personal ethics allow me to work on developing battlefield weapons, but not on weapons of mass destruction. Years ago, an employer tried to shift me from work on antiaircraft missiles to long-range ground-to-ground missiles. I said that I would not feel right about working on weapons of mass destruction that might be used against cities. My employer informed me that this was the only work available at that time, as events beyond his control had eliminated the other work. I therefore resigned rather than accept the new assignment.[25]

The offensive/defensive weapons distinction has also occasionally formed the basis for institutional research policies. For example, Ohio State University, which accepted over $5 million in defense contracts research in 1982, has adopted a policy that bans work on offensive weapons.[26] Other universities, such as MIT, do not attempt to evaluate the end-uses of the research projects accepted by its faculty, but have adopted policies regulating the procedural aspects of military research, opposing the conduct of classified research on campus, restrictions on publication or participation by foreign students, and so forth.

Some individuals seeking to resolve the conflicting ethical issues presented by military research conclude that drawing lines on the conduct of such work is a hopelessly useless task because of the uncertainties associated with the applications of the research and the political process. Some would argue that it is more effective to work for political change in the policies that result in objectionable research projects (such as the Strategic Defense Initiative) than to refuse assignments to such research.

One researcher at the Stanford Research Institute (SRI) adopted this approach in the midst of the Vietnam controversy:

> Say I were in a position to decline research from the Department of Defense that had to do with Viet Nam. I would prefer, if only *I* were involved, to reject it, because I don't personally believe in the government's policy on Viet Nam. But I couldn't allow myself to do that for moral reasons, because I will have to assume that the public wants the war in Viet Nam to go on, because the Congress has authorized money for it and this agency of government is doing its duty by asking SRI, [which] is competent to do this research, to do it. So, I should accept this contract . . . because I believe that the government reflects the will of the public. Now the public may be a little misguided, in my opinion. But the way for me to do something about it is through the channels which our founding fathers have established. . . . But I should not subvert the public will. I should use the existing channels by writing, by talking, by educating people, and so on, to vote differently.[27]

Reconciling these various positions requires a broader moral framework that invokes the just war theories, the issues of wrongful orders, the role of the citizen in the political process, and so forth, a task that goes far beyond the limits of this discussion. Two themes have emerged, however, and they deserve more attention:

1. The absence of a shared framework among scientists for evaluating the moral issues associated with individual involvement in military research.
2. The presence of factors that might provoke a search for a common conscience on this subject.

In writing about issues of social responsibility in science, Sir Solly Zuckerman has commented on the absence of universal standards for judging the end-uses or purposes of scientific work:

> When one talks about the social responsibility of scientists, it is thus carrying naivete to the extreme to suppose that they speak with one voice and that they share a common conscience when it comes to the application of scientific knowledge.[28]

Other writers, however, have suggested that the shared values forming the basis for scientific conduct could or should be extended to include general agreement regarding the purposes or ends to which science should be directed. Because scientists share common methods and values in their professional activities, because they are said to be grounded in reason rather than passion or ideology, it has been suggested that they, perhaps more than others, should be able to agree upon the directions in which they would like to see their knowledge applied, or at a minimum, restrict activities that

British physicist John Ziman has termed "wicked" science. In some cases it has even been suggested that scientists should put their allegiance to science itself above all other loyalties in supporting selected political causes.

For example, in 1941 British historian J. G. Crowther recommended a set of eight social responsibilities for scientists. He urged scientists to consider the social implications of their work, and to become politically active in order to contribute to the development of constructive social forces regulating the conduct of research. His final recommendation was that in war, scientists should "consider which side was the less inimical to science, and then do what was possible to see that it was not defeated."[29]

Another example of the difficulty in forming a universal framework for a common conscience among scientists can be found in the correspondence between Norbert Wiener and Louis Ridenour in the late 1940s.

In early 1947, after the bombings of Hiroshima and Nagasaki, MIT mathematician Norbert Wiener published a letter in the *Atlantic Monthly* in which he announced his intention to restrict and limit his scientific communication on the basis of a decision not to "publish any future work . . . which may do damage in the hands of irresponsible militarists."[30] Wiener, who had worked on guided missiles during the war, described how he reacted to a request that he provide a copy of his research paper to a scientist at a major aircraft corporation:

> The policy of the government . . . during and after the war had made it clear that to provide scientific information is not a necessarily innocent act, and may entail the gravest consequences. . . . It is perfectly clear also that to disseminate information about a weapon in the present state of our civilization is to make it practically certain that the weapon will be used. . . . If therefore I do not desire to participate in the bombing or poisoning of defenseless peoples—and I most certainly do not—I must take a serious responsibility as to those to whom I disclose my scientific ideas.[30]

Wiener's letter generated many supportive responses within and outside the scientific community. It also provoked a reply that appeared in a later issue of the *Atlantic Monthly*.

In an article titled "A Scientist Fights for Peace," University of Pennsylvania physicist Louis Ridenour, who had worked on the development of radar in the war, challenged Wiener's claims. Ridenour indicated that his disagreement with Wiener rested on two points: his interpretation of the social responsibility of the scientist, and the best way of implementing this responsibility. With respect to the first point, Ridenour wrote the following:

> Wiener clearly believes that the scientist is the armorer of modern war, and as such holds a responsibility of unique importance. I feel that the social responsibility of the scientist is unique in no way. It is identical with the social responsibility of every other thinking man, except for one special and temporary thing. It is necessary today to educate the nonscientific public [about atomic energy] so that all the people can participate in the decisions they will have to make concerning the organization of society in such a form that wars become less likely.[31]

On the second point, Ridenour was more critical. He argued that Wiener was, in effect, doing everything he could to make American preparations for war ineffective. This, Ridenour argued, was sabotage:

> I regard it as deplorable that our nation is preparing for war, and I prefer to leave to others the actual work involved; but so long as it is the policy of our nation to prepare for war, I shall certainly not attempt to impede such preparations. . . . I conceive the duty of the peace-lover to be that of working for a world in which national arms are no longer desired by a majority of the people of this country or of the world.[31]

These vignettes and others suggest that scientists, like other citizens, do not agree on what the ultimate purposes of science should be in the context of military work. As noted, some individuals firmly believe that scientists should not work on military projects, particularly projects involving the design or development of nuclear weapons. This line of reasoning can be traced, in part, back to the writings of Sir Francis Bacon, who wrote about the standards to be observed in his House of Solomon:

> And this we also do: we have consultations [amongst ourselves] which of the inventions and experiences which we have discovered shall be published and which not, and take an oath of secrecy for the concealing of those which we think fit to keep secret, though some of those we do reveal sometimes to the state and some not.[32]

Others, like Ridenour and many scientists associated with arms control as well as arms development efforts, argue that it is ultimately the public that must decide which national policies should guide government-funded research and development, and that the scientist should participate in, but not preempt, this public decision-making process. This group also has its antecedents in the early history of science. Still others believe that the scientist should simply seek to do good science, and leave the matters of ends, uses, and consequences, to others, suggesting in some cases that even if individual scientists refuse to develop lethal military projects, others will do it in their place.

Regardless of their own personal views and decisions, most scientists and engineers probably agree that the matter of deciding whether to work on military research in general or on particular research projects is an individual rather than a collective problem. Even in the absence of any common conscience or universal framework that might be used as a standard for evaluating selected research projects, there does appear to be strong support for the right to make individual choices.

This consensus for the concept of individualism appears in a response to the 1970 SESPA pledge, noted earlier. MIT physicist Lee Grodzins, who chaired the then newly organized Union of Concerned Scientists (UCS), wrote in a letter to *Science*:

UCS has taken no position on the SESPA pledge not to participate in war research. We do not intend to. . . . The UCS members . . . held the view that signing such a pledge is matter for personal conscience, not collective intimidation. . . .
There are circumstances when some of us would work on weaponry. We are convinced that now is not such a time. We devote our energies and our talents so that the time may never come.[33]

SEEKING A COLLECTIVE RESPONSE

Restrictions on individual choice, created either by economic pressures, institutional policies, or other forms of social limitations, may generate pressures within the scientific community to develop broader collective responses to the problem of drawing lines.

In times when economic conditions support a wide range of vocational choices and alternatives, the ethical issues embedded in drawing the line between certain categories of military work, or in choosing between military and civilian research, often remain unexplored and unresolved. Each individual judges the appropriateness of research assignments according to the technical fascination of the research, economic incentives, or personal moral criteria. When alternatives are limited, however, when some individuals feel forced to consider projects they find morally objectionable, the demand for collective responses by the profession becomes more immediate and controversial. The tensions generated by the trade-offs involved in pitting one's conscience against economic realities stimulate broader searches for solutions to relieve the personal stresses provoked by the constraints scientists face. These solutions include proposing policies to promote public investment in nonmilitary research programs (the conversion or swords-into-plowshares approach); calls to scientists and engineers to become more involved in public education and political lobbying to change the conditions that have limited their options; or withdrawal from a field of science in order to seek opportunities that place less stress on one's moral convictions.

Such restricting times stimulate arguments that scientists need to address and influence in some collective fashion the social consequences of their work. Such times also provoke counterarguments, urging scientists to stick to science and to leave politics in the hands of government authorities.

Almost two decades ago, political scientist Joseph Haberer observed that scientists had failed "to define, [to] delineate, or even to recognize the nature of the problem of responsibility," and he strongly recommended that there was a real need "to consider and clarify the problem of social responsibility as it applies to science."[32] I suggest that such definition, delineation, and even formal recognition of the problem of responsibility will remain a latent issue unless stimulated by external constraints on individual choices by members of the scientific and technical enterprise. If these constraints arise, they will provoke a search for a collective response to the problem of social responsibility of science in the form of political or social change.

AREAS OF POSSIBLE CHANGE

It is useful now to consider areas in anticipation of a future search for collective solutions to the problem of social responsibility in military research.

1. The problem of drawing lines needs to be recognized as a historical problem, and the scientific community needs to participate in a broader social process of resolving moral conflicts associated with military service. Just as scientists have always worked on military problems, they have also always raised concerns about the military applications of scientific and technical knowledge.

   It would be useful to know more about the social conditions that cause scientists to turn their attention to problems of war rather than to the concerns of how to keep peace. Is this simply a matter of economic incentive? Is there something fundamental to the nature of science that makes it better suited as a tool of military systems than as a resource for the peaceful resolution of political conflict? In his book *The Pursuit of Power*, historian William McNeill notes the development of early cannon technology required more than a century of stubborn commitment to the search for the principles of explosives, in a time when there was little likelihood that such experimentation would be successful and when there was little financial support for such experimentation—the economics of catapults as a means of attack were much more attractive.[34]

   What was it that drove these scientists of the medieval age into the field of explosives? McNeill suggests that it may have been a deeply psychological motive: a primitive fascination with sudden changes in heat and light. The same primitive psychological forces may have been at work in the Manhattan Project.

   Can this psychology be directed toward the development of devices to manage conflict in a more peaceful manner? Is there a technology that could absorb aggressive attacks so that they become meaningless? What technologies are suitable for keeping peace rather than making war?

2. Another approach to consider is the development of a frame of reference for "unlawful science." International agreements have led to military codes that provide a basis for a soldier to refuse an unlawful military command. In seeking ways to implement a broader social responsibility in science, one avenue might be to develop a list of weapons that are banned by international treaty and to formulate an international protocol urging scientists of every nation to refuse assignments that directly contribute to the development of such weapons.

   The risk here is that such refusal might lead to a charge of treason or sabotage by the government so challenged and to punishment for the individual scientist. To provide protection against such risks, or-

ganizations such as the International Council of Scientific Unions (ICSU) might consider maintaining records of government compliance with international treaties involving scientific and technical issues. Rather than placing the burden on the individual to challenge violations when they occur, ICSU might censure governments found to be in violation of international agreements in the same way that they prohibit conferences in nations that refuse to comply with their standards of scientific freedom.
3. In the education of young scientists and engineers, historical and critical examinations of the relationships between science and the military should be encouraged. Students should be informed early in their training of some of the ethical dilemmas raised in the course of military research. They should be asked to consider the likelihood that they will be able to work on projects that are consistent with their own moral standards, and they should be provided with examples illustrating alternative approaches to ethical dilemmas in military research.
4. Finally, legal and management experts should consider whether conscientious objectors in the workplace merit legal protection. Our society has a tradition of recognizing the rights of those who refuse to bear arms in service to their country. I believe an argument can be developed suggesting that a similar right should extend to those who refuse to *make* arms. The right to work, and in particular, the right to be free from arbitrary dismissal without just cause, is a legal concept that is evolving in various state and federal courts.[35] The issue is fraught with conflicting rights, complexity, and uncertainty. But it requires scholarly examination and discussion to identify the points where there is a need for institutional policies and perhaps legislative changes to safeguard the right of individual choice.

In conclusion, the problem of drawing lines in military research will continue to be resolved through individual choice in the absence of a common conscience about the ethical issues raised by military research. Yet there are steps that can be taken now to support and inform individual choice that will provide the broader scientific community an opportunity to address more directly the issue of social responsibility in military research.

REFERENCES

1. MUSTE, A. J. Letter to A. Einstein dated September 15, 1947. Located at: Records of the Fellowship of Reconciliation (FOR), Swarthmore College Peace Collection (SCPC), Swarthmore, PA.
2. For more details of Muste's life, see ROBINSON, J. A. O. 1981. Abraham Went Out. Temple University Press. Philadelphia, PA.
3. EINSTEIN, A. Letter to A. J. Muste dated October 11, 1947. Located at: FOR, SCPC, Swarthmore, PA.

4. WITTNER, L. S. 1969. Rebels against War: The American Peace Movement, 1941-1960 (quote from letter): 177. Columbia University Press. New York, NY.
5. WITTNER, L. S. 1969. Rebels against War: 177.
6. UREY, H. C. Letter to A. J. Muste dated June 10, 1946. Located at: FOR, SCPC, Swarthmore, PA.
7. BETHE, H. Letter to A. J. Muste dated December 16, 1946. Located at: FOR, SCPC, Swarthmore, PA.
8. HIGINBOTHAM, W. A. Letter to A. J. Muste dated August 5, 1946. Located at: FOR, SCPC, Swarthmore, PA.
9. For a short history on the origins of conscientious objection, see WALZER, M. 1970. Obligations: Essays on Disobedience, War and Citizenship. Harvard University Press. Cambridge, MA. See also SCHLISSEL, L., Ed. 1968. Conscience in America. E. P. Dutton. New York, NY.
10. WALTZER, M. 1970. Obligations: 125.
11. Kapitza's decision is documented in several sources, including PARRY, A. 1968. Peter Kapitsa on Life and Science. Macmillan. New York, NY. See also HOLLOWAY, D. 1986. Life of a scientist (book review). Science 232: 1559.
12. Wiener describes his ethical positions in his autobiography; see WIENER, N. 1956. Moral problems of a scientist. In I Am A Mathematician: Chapter 14. MIT Press. Cambridge, MA.
13. Paschkis's friendship with Muste and the origins of the SSRS are described in ROBINSON, J. A. O. 1981. Abraham Went Out: 143-144.
14. A discussion of Einstein's views on pacifism is included in NATHAN, O. & H. NORDEN. 1960. Einstein on Peace. Simon & Schuster. New York, NY. See also CLARK, R. W. 1971. Einstein: The Life and Times. Avon. New York, NY.
15. Declaration of the German Nuclear Physicists. 1957. Bull. At. Sci. 13: 228, 283-286.
16. ROBINSON, J. A. O. 1981. Abraham Went Out: 143-144.
17. Scientists and Engineers for Social and Political Action. 1971. Science 171: 156.
18. For a description of da Vinci's action, see NEF, J. U. 1950. War and Human Progress. Harvard University Press. Cambridge, MA. See also MUMFORD, L. 1979. Interpretations and Forecasts: 1922-1972: 315. Harcourt Brace Jovanovich. New York, NY.
19. NEF, J. U. 1950. War and Human Progress: 121-122.
20. NEF, J. U. 1950. War and Human Progress: 196-197.
21. SZENT-GYORGYI, A. 1963. Science, Ethics and Politics: 55. Vantage Press. New York, NY.
22. STRAUSS, L. L. 1962. Men and Decisions: 406. Doubleday. Garden City, NY.
23. BAXTER, J. P., III. 1946. Scientists against Time: 3-4. Boston, MA.
24. BLUMBERG, S. A. & G. OWENS. 1976. Energy and Conflict: The Life and Times of Edward Teller: 240-241. G. P. Putnam's Sons. New York, NY.
25. Letter. 1975. IEEE Spectrum. March: 24.
26. Article. 1983. Kansas City Times. January 13: p. A7.
27. SCHEVITZ, J. M. 1979. The Weaponsmakers: 136. Schenkman. Cambridge, MA.
28. ZUCKERMAN, S. 1970. Beyond the Ivory Tower: 4. Taplinger. New York, NY.
29. CROWTHER, J. G. 1967. The Social Relations of Science: 450-451. Crescent. London.
30. WIENER, N. 1947. A scientist rebels. Atl. Mon. 179(1): 46.
31. RIDENOUR, L. 1947. A scientist fights for peace. Atl. Mon. 179(5).
32. Quoted in HABERER, J. 1969. Politics and the Community of Science: 44. Van Nostrand Reinhold. New York, NY.
33. GRODZINS, L. 1971. Letter. Science. April 16: 214, 216.
34. MCNEILL, W. H. 1982. The Pursuit of Power: 83. University of Chicago Press. Chicago, IL.
35. For a thorough discussion of the laws that protect employees against unjust discharge, see KOHN, S. M. & M. D. KOHN. 1988. The Labor Lawyer's Guide to the Rights and Responsibilities of Employee Whistleblowers. Quorum. New York, NY.

The Fine Structure of Military Research as an Essential Background for Discussions of Ethics
Remarks on a New Liberation Ethics

RUSTUM ROY
Science, Technology, and Society Program
Pennsylvania State University
University Park, Pennsylvania 16802

INTRODUCTION

Arguments over what conditions justify military research often parallel arguments over what conditions justify war. The former often suffer, however, because the term military research, which covers an enormous range of activities, is often used much too loosely. In this paper, we will distinguish between different kinds of military research, and we will consider how the term military research, if used indiscriminately, could confuse ethical issues.

One source of confusion concerns supposed parallels between the ethics of war and the ethics of military research. Using just war theory, which emphasizes the collective nature of war, one may argue that it is possible to distinguish between just and unjust wars. An analogous theory for military research, however, could prove to be inadequate. For example, the personal and collective issues raised by military research may not correspond to the personal and collective issues raised by war. Furthermore, if one is unable to distinguish between different kinds of military research, how can one account for the possibility that these issues may require different treatments depending on the kind of military research in question?

SUPPORT FOR MILITARY RESEARCH

As a director of research in the physical sciences and engineering, I am familiar with the institutions that support military research. These institutions are referred to throughout this paper as defense-related organizations (DROs).

I have chosen to use the term DRO-supported research as a means of describing the inclusive whole of what is often called military research. By referring to the support of military research, financial or otherwise, we can objectively define the boundaries of the field. Unfortunately, the simpler term

DOD-supported research will not do because in the United States very large amounts of military research are outside the Department of Defense (DOD). Nuclear weapons, for example, are developed by the Department of Energy (DOE). To some extent, the National Aeronautics and Space Administration, the Central Intelligence Agency, the National Security Council, the Arms Control and Disarmament Agency, the Federal Emergency Management Agency, and private corporations (albeit to a remarkably small extent) all provide support for military research. Parts of these agencies may be considered DROs, although for statistical purposes DOD and DOE weapons cover 95% of the effort.

In general, the budgets and broad range of research activities of the federal agencies named above are public knowledge (always excepting, of course, the research components of the very large so-called black budgets, which have been estimated to amount to $10 billion). For fiscal year 1989, the public budgets amounted to roughly 75% of the total research and development effort of the U.S. government; that is, these budgets totaled about $64 billion between 1988 and 1989. FIGURE 1 gives some details on the size of

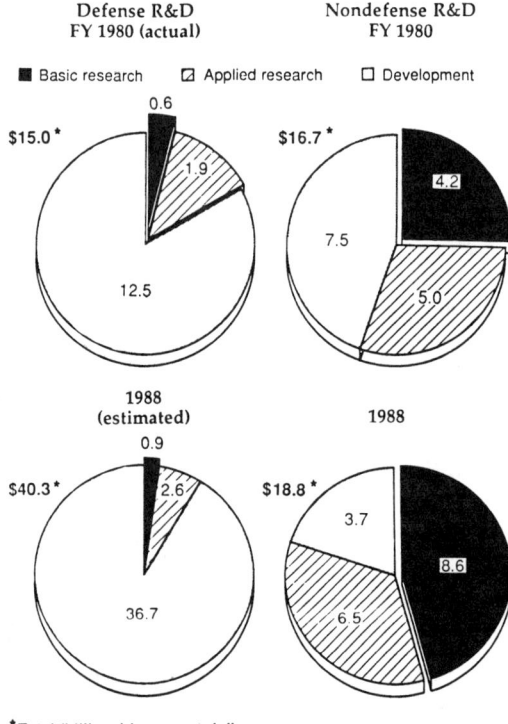

FIGURE 1. Federal research and development budgets by function. Please note that the values indicated for defense spending may not include all spending by DROs.

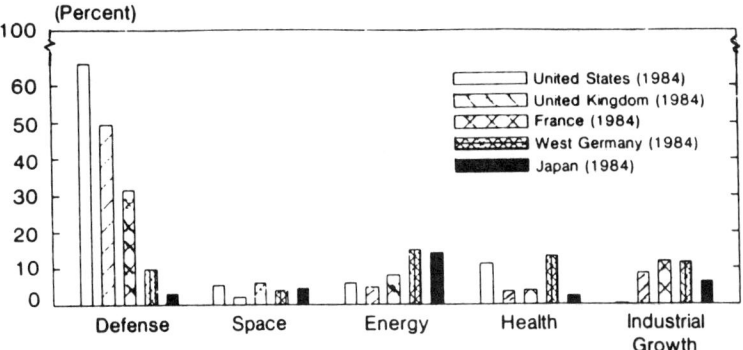

FIGURE 2. Distribution of government research and development expenditures among selected objectives by country. The percentages on the vertical axis correspond to the heights of the bars for U.S. expenditures. The bar heights for the other countries, however, are in proportion to U.S. bar heights as the expenditures by the other countries are in proportion to U.S. expenditures. Note the radical differences. These differences reflect the gross overemphasis on military research in the United States, which is the key ethical question for the science community.

the domain that is the subject of this inquiry. About one-third of the engineers and scientists in the nation are supported (many, of course, part-time) by DROs. The ethical issues are clearly secondary to many other factors (for example, economics, personal advancement, and patriotism) in determining whether individuals will participate in DRO-supported research. FIGURE 2 shows the research and development effort of several nations arranged by objectives. Extraordinary differences between nations may be observed. The United States and the United Kingdom lead in DRO-supported research by an enormous margin. The negative correlation of these high percentages for DRO-supported research with increases in productivity is well established, and will be referred to later. These differences cannot be correlated principally with the personal ethical choices of participants in the different countries. Rather, participation of a sufficient number of scientists and engineers in *any* research (including, say, Nazi hypothermia studies) in any country seems to be assured provided the funds are available. Thus the very low percentage of Japan's research and development expenditures devoted to defense cannot be used as evidence of superior or different ethical standards.

RANGE OF ACTIVITIES SUPPORTED BY DROs

Military research, or at any rate research supported by DROs, covers a wide range of activities. Some activities supported by DROs seem to have little, if any, military use. Other activities, however, have obvious military use. Given that military applications are apparent in different degrees for different activities, we may order different kinds of DRO-supported research from those with the least obvious military use to those with the most obvious military use (or from those least likely to be classified to those most likely to be classified). I have found that the following five categories are useful:

1. **Atelestic Research.** By atelestic I mean to refer to those activities usually thought of as being typical of basic research. Atelestic research, which enjoys totally open access, need not be carried out with any particular application, military or civilian, in mind. (Examples: plate tectonics and new superconductors.)
2. **Telestic Research.** By telestic I mean to refer to basic and applied research in which both military and civilian applications may be apparent, but in which the military applications are so distant or tenuous that questions about classification never arise. (Example: new transducers for both medical ultrasonics and submarine detection.)
3. **Basic and Applied Research on Militarily Relevant Topics.** Research of this kind is potentially so valuable to the military that classification or some other restriction may be imposed, even if civilian applications are also apparent. (Example: microwave absorbers for coating stealth bombers.)
4. **Defensive Equipment Research.** This category includes research on truly defensive military hardware, that is, hardware with no potential for the direct killing of humans. Such research is always classified. (Example: ceramic body armor.)
5. **Offensive Weapons Research.** This category includes research on the instruments used directly for killing humans and destroying property. Such research is always classified. (Examples: smart bombs and nerve gas.)

Some elaboration on each of these is in order.

Atelestic Research

The system of widespread research in the physical sciences and engineering, particularly at universities, started very soon after World War II. Before World War II, the worldwide research and development effort, including that in the United States, was very largely independent of DROs, and participation in military research could in fact be largely a conscious ethical decision.

In *Lost at the Frontier*,[1] a book I coauthored, the events leading to the early involvement of the DOD in supporting university research, including the early entry by the Office of Naval Research (ONR), are described by several different persons. The failure to create Vannevar Bush's National Research Foundation in the early Truman years meant that the ONR and the National Institutes of Health (NIH) were able to assume the responsibility of funding most of the physical and biological research at the universities. Even when the National Science Foundation (NSF) was finally created in

1950, it received much, much smaller appropriations than a grateful nation and the Congress were willing to give the DOD for those wonder workers who were credited with delivering the nation from Germany and Japan—the scientists (but, absurdly, not the engineers, who had done the real work). Thus, in America, much general basic research came to be supported by the DOD. This idiosyncratic situation may have reached the zenith of absurdity when, in response to Soviet advances in space, Congress met the alleged problem of insufficient science and engineering education by passing the National Defense Education Act (NDEA). Thousands of doctorates—even in the social sciences—were granted to NDEA fellows.

If, after the war, support for basic science, especially university science, was to be maintained at the levels to which the leadership had become accustomed in the glory days of the Manhattan Project, those who would justify maintaining this support would have to employ at least a minor subterfuge. In 1946, the DOD and its point subgroup, the ONR, did indeed work up a justification for truly atelestic (basic) research: such research was defined as a necessary precondition for a healthy national technological base, and this base was linked, through a series of weak arguments, to the defense needs of the United States.

A historical (and very personal) anecdote should suffice to illustrate how broad a mission the DOD assumed. On July 1, 1948, Pennsylvania State University received its first DOD grant. This grant, which was one of the very first received by the university's geophysics section, was given for the study of the fundamentals of mineral and rock metamorphism in the earth. My contribution to this military research—I was the first person employed on this first defense grant—consisted of studying the stability of minerals in the earth.

After 1950, with the NSF in place, the arguments linking basic research and the nation's defense should have withered. They did not. The Congress simply was used to giving large sums to the DOD. Even the trickle down to the ONR was large compared to the budget for the NSF, and more and more basic research began to be funded by various mission agencies. NASA, in particular, contributed to this unfortunate trend. The science community meanwhile began to like the idea of multiple sources of funding for any number of different kinds of research. Not only did it mean much more money than they could ever get under one agency, but the idea of having multiple sources became one of the truly great U.S. inventions in science-funding strategy. The fact that someone, somewhere up the line, had to approve this very basic, wholly open research as being of importance to the DOD mission did not concern individual scientists in the slightest.

Let us take a close look at the characteristics of this particular category of DRO-supported research:

1. This category covers research that could be, and in fact is, also supported by many non-DRO agencies. In fact, the same work often is necessarily funded by DRO and non-DRO agencies. Often the same

individual switches the support for the same work from a DRO to a non-DRO.
2. Research is totally unclassified. There are absolutely no restrictions on who does it or on where results are reported. No permissions are required at any time to disseminate the results.
3. Research need not have any particular application as a goal—whether civilian or defense related.

Telestic Research

This is an under-used category in science policy circles; lack of familiarity with this category causes much confusion therein. The Mansfield Amendment, which was passed in 1969 during the Vietnam War, forbade the DOD from supporting any (basic) research that was not in some way relevant to its own mission. Many DOD projects and areas of research were indeed dropped over the next few years, but virtually any research the DOD project manager wanted to save could be justified somehow. Over the years, the Mansfield Amendment has disappeared from view, and DROs again all support virtually any research in any field, but obviously spend most of their money in fields where the following criteria apply:

1. Same as the first criterion for atelestic research.
2. Same as the second criterion for atelestic research.
3. The research should be part of a chain in which a manager's justification statement links the research to a particular application of interest to the DOD.

Note that although this basic information has value to the DOD, it has so little specific value that absolutely no effort is made to restrict the information.

I believe the criteria of openness and complete access (even to Soviet nationals) and the interchangeability of supporting agencies make atelestic and telestic research moot when discussing military research.

Basic and Applied Research on Militarily Relevant Topics

This category of research begins to approach what generally seems to pass as a description of military research. The characteristics of research in this category can be described as follows:

1. Basic or applied research specifically and closely relevant to some military objective.
2. The results of the research should be kept from the general scientific community (that is, permission must be granted to present or publish). No aliens should have any access to the information.

3. Formal military classification may be instituted, with locked files and drawers, for example, to prevent even colleagues from seeing the results.

An example of this kind of research in the news is the stealth bomber, a new warplane that is supposed to be invisible to radar. One of the ways to make the plane invisible to radar is to coat it with a special film or paint that will absorb radar waves very effectively (instead of reflecting them to an enemy's detectors). Clearly the DOD would not want the funded university researcher to publish a breakthrough in this field in a Soviet journal, nor would it want a visiting Eastern European scientist to sit next to the person doing the work. The degree to which information is restricted indicates the degree to which it has military relevance.

In deciding whether to restrict information, however, one must frequently face a dilemma. Either one eschews restrictions, and accepts a risk to security, or one imposes restrictions, and accepts a risk to economic development. For example, radar-absorbing coatings developed for the stealth bomber could be useful on warplanes, but they could also be useful in microwave ovens. Restricting information for security reasons could hurt one's own industry.

Defensive Equipment Research

Research in this category is often of an applied or developmental nature. The objectives of this research are immediately, unquestionably, and exclusively of value to the DOD.

The characteristics of research in this category are as follows:

1. Research is definitely classified because it has direct and immediate links to DOD applications, and because enemies could use it to our disadvantage.
2. It has no potential for directly killing human beings.

An example of this kind of research is the development of ceramic body armor (like bulletproof vests). There is no way in which ceramic body armor can be even misused to hurt other persons, but its use could make our personnel safer and hence more able to kill.

So-called defensive weapons are borderline cases between this category and the next (and last) one, but insofar as they can be easily used to inflict death and destruction, I would place them in the last category.

Offensive Weapons Research

Finally we come to the least ambiguous category of military research, research on weapons—specifically, research on weapons that can be used to kill people in small or large numbers.

Here the ethical issues may be most clearly related to classical positions. Should a scientist devote his or her talents to improving the means of killing people on the "other side"? Is there, in addition, some special moral compunction that should apply to perfecting weapons of mass destruction (that is, nuclear, chemical, and biological weapons)?

In this section—that is, in the five categories outlined above—I have tried to present the insider's view of the great, big, fuzzy ball of human activities that goes under the term DRO-supported research. It should be obvious that no ethical principles can be developed that would apply equally to all five categories. Those unfamiliar with research and development policies and practices in the United States can perhaps sharpen their arguments by addressing the different categories separately.

I believe that the background provided in this section is absolutely indispensable in carrying out any worthwhile discussion of military research. Most readers would probably agree that lumping together atelestic research (say, on plate tectonics) and offensive weapons research (say, on hydrogen bombs having high neutron yields) would be futile in any attempt to resolve ethical dilemmas in our age. For most people unfamiliar with DRO-supported research, however, the term "military research" applies to only the last two of our five categories—defensive and offensive weapons research. If we could all recognize the reality that huge DRO expenditures by the United States support research very different from "military research" (as perceived by the public), the ethical debate would be greatly improved. Of course, this does not by any means eliminate debate on the very different ethical (and tactical) question as to whether the United States should support general, basic research—such as that described in the first three categories—under the auspices of the Department of Defense.

LIBERATION ETHICS FOR THE SCIENCE COMMUNITY

The personal ethics of participating in a collective group in making war are most often argued on the basis of the collective justice of the war. If the war is just, personal participation, even sacrificing the personal ethic against killing, is justified. Of course the pacifist religious communities argue a more personalistic ethic, which really has little to do with the social ethic of war, and is instead concerned with individuals killing other individuals.

I believe that more attention should be devoted to ethical issues often missed in connection with DRO-supported research because they are not personal but group ethics. Liberation theologians have been at great pains to remind the Western Judaeo-Christian community that a great deal of the ethics in the Old and New Testaments applies to the "nation" (or tribe or group) and not to the individual. Their exegesis of the twenty-fifth chapter of Matthew presents a clear case where "all the nations" are called before a great assize.

Analogously we must now ask whether there are collective social-ethical questions with which the science and engineering community should be confronted. I believe so. Furthermore, I believe that some of these are profoundly important because they are so intimately connected to science's absolute commitment to truth and precision.

These relations between personal and societal ethics of war and research areas are (crudely) illustrated in FIGURE 3, which attempts to draw the distinction that in war itself individuals do not become involved as individuals, only as part of a collective (tribal or national) group. Participation in wars may be defended as ethical on the basis of traditional just war theories, which of course come under attack when civilian populations are threatened. In re-

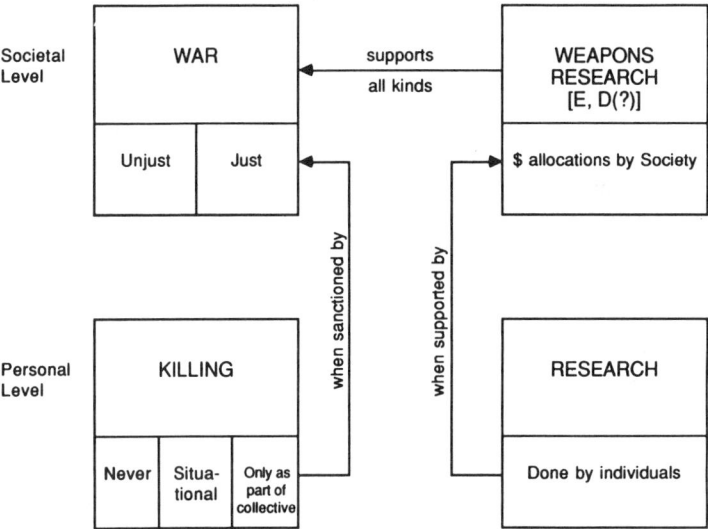

FIGURE 3. A schematic showing some differences between participating in war and participating in weapons research. Individuals take part in any particular war as part of a collective that has judged the war to be just; individuals do weapons research as individuals, with the only collective ethical decision being one of how to allocate research and development resources.

search, however, one may even develop weapons without having any relation to a societal decision on whether a particular war is just or unjust. One's research could eventually be used for either a just or an unjust war. Of course, if a war were going on that one perceived to be unjust, one might refuse to carry out one's research, but in most instances one contributes less to particular decisions about actions in the present, and more to decisions about the kinds of options that will most likely be resorted to in the future. These latter decisions are less dramatic but perhaps more important because they are made regularly—and frequently. Specifically, they have to do with the budgets allocated to different kinds of research. Budgetary decisions raise ethical issues that should be given at least as much scrutiny as issues about just and unjust wars. Unfortunately, the problems of deciding on budgets and allocating resources are much neglected as topics in social ethics.

I believe that by far the most important ethical issue that should engage our attention is the question of the general domination of the scientific enterprise by the military machine. Is there a real community or society of scientists and engineers? If so, does it have any common ethics? If so, what fraction of its effort would the scientific community choose to devote to research and development for the military? Some might find it ominous that a third of the national research and development personnel is devoted to defense-related work. What may be even more debilitating to the moral fiber of collective science, however, is that the direction of science in the post-World War II era in the West and in the Soviet Block (but not in Japan) has been defined not by some process intrinsic to the development of science (if we may say this process does indeed exist), but nearly always by the military establishment, which is always willing to pay for the most exotic or esoteric science. And the community's definition of "frontier" or "high-powered" science has been so configured that narrow, machine-intensive topics with no impact on neighboring fields, and no reference to the human scale or to human need, have received the bulk of research. What a far cry from Einstein's plea at the California Institute of Technology in 1931:

> It is not enough that you should understand about applied science in order that your work may increase man's blessings. Concern for man himself and his fate must always form the chief interest in all technical endeavors, concern for the great unsolved problems of the organization of labor and the distribution of goods—in order that the creations of our minds shall be a blessing and not a curse to mankind. Never forget this in the midst of your diagrams and equations.[2]

In deciding what constitutes good (or the best) science, we could perhaps keep in mind Percy Bridgman's learnings from quantum mechanics:

> Finally, I come to what it seems to me may well be . . . the most revolutionary of the insights to be derived from our recent experiences in physics, more revolutionary than the insights afforded by the discoveries of Galileo and Newton, or of Darwin. This is the insight that it is impossible to transcend the human reference point. . . . The new insight comes from a realization that the structure of nature may eventually be such that our processes of thought do not correspond to it sufficiently to permit us to think about it all.[3]

The ranking of different kinds of research according to their value to humanity has occupied many scientists and engineers. Alvin Weinberg, for example, has suggested that the "impact on neighboring fields" be a criterion for judging science.[4,5] Suggestions offered by scientists, however, have been all but ignored in the push for budgets with more defense-oriented research and development.

It is ironic, perhaps, that the domination by Japan of a radically different approach to science and technology will, in the next two or three decades, do more to liberate Western science from its subservience to military prerogatives than any ethical considerations.

Another important ethical issue facing the scientific community is the question of how much value scientists and engineers should place on research as opposed to teaching. As Shapley and Roy have shown in detail, the decline of science education in the United States can be laid squarely at the feet of the scientific community itself.[1] It was the body most directly representative of the scientific community, the National Science Board, that systematically and monotonically reduced the proportion of the NSF budget devoted to teaching from nearly 50% in the late fifties to essentially 0% in the early Reagan years. If scientists show so clearly that they believe research is vastly more important than the lowly task of educating American citizens about science, the public can hardly be expected to have much enthusiasm for learning about science.

One last ethical issue concerns the scientific community's failure to grapple with the values intrinsic to its present collective processes. One of these processes rewards narrow differentiation and specialization and penalizes integration and breadth. Another rewards abstraction and ignores personal and societal issues. It is interesting to observe the nature of the so-called Alternative Nobel Prizes awarded in Stockholm by the Right Livelihood Foundation every year, a week before the Nobel Prizes are awarded. It has many of the characteristics noted above as being missing from the goals of contemporary Western science. The hegemony of the post-World War II model of science, however, is coming to an end. Indeed, some observers, such as V. Weisskopf, former Director General of CERN, believe that the expansion of knowledge usable in the everyday world is rapidly approaching an asymptote.[2] It is an opportune time for a new set of values to be adopted for world science. I have suggested elsewhere that the most basic and most valued science be defined as that advancement of knowledge which makes the most beneficent and widespread impact on humanity.

REFERENCES

1. SHAPLEY, D. & R. ROY. 1985. Lost at the Frontier. ISI Press. Philadelphia, PA.
2. ROY, R. 1982. Experimenting with Truth. Pergamon Press. New York, NY.
3. BRIDGMAN, P. W. 1955. Reflections of a Physicist. Philosophical Library. New York, NY.
4. WEINBERG, A. 1963. Criteria for Scientific Choice. Vol. 1(2). Minerva. New York, NY.
5. WEINBERG, A. 1964. Criteria for Scientific Choice. Vol. 2(3). Minerva. New York, NY.

Science Pure and Applied as Essentially Involved with Ethics

KRISTIN S. SHRADER-FRECHETTE

Department of Philosophy
University of South Florida
Tampa, Florida 33620

INTRODUCTION

Virtually all scientists and philosophers hold the positivist position that science, in itself, is not laden with ethical values, and that purely scientific or cognitive decision-making is not subject to moral appraisal. This is allegedly because theoretical science itself (apart from the practical ends to which it may be applied) involves only *epistemic* assessment, not calculation of the utilities (for various agents) that are associated with different theoretical outcomes.[1]

Despite the predominance of this positivist view, I contend that a scientist might be morally praiseworthy or blameworthy for deciding that something is or is not the case. Admittedly, however, not every act of every scientist is essentially moral. There are at least three arguments suggesting that pure science does involve judgments about ethical values. After discussing these arguments, I consider possible objections to my position, especially as formulated by persons like James and Alston. I close by suggesting some of the practical consequences that follow, for scientific research, if pure science indeed involves an ethics of belief.

ARGUMENTS FOR ETHICAL JUDGMENTS IN PURE SCIENCE

In every area of pure and applied science, scientists have a duty to be open-minded in assessing novel, controversial, or innovative scientific results. The existence of such a duty is evident because, did it not exist, researchers could not be said to be wrong or to be culpable for their closed-mindedness or their biased judgments, particularly when they evaluated claims threatening the status of their own research. But we do hold scientists accountable for their biases, and therefore there must be a duty not to be closed-minded.

Moreover, the duty, and scientists' value judgments about it, appear to be distinctively ethical, for at least three reasons. 1) The duty is closely akin to the obligation not to obstruct fair play or equal consideration of all scientists' interests. Such an obligation I take to be self-evident. 2) The duty not to be closed-minded also reflects an ethical value because the consequences of recognizing (or not recognizing) it have effects on the welfare of others, for example, the proponents of the innovative scientific theories, and perhaps on their opponents as well. In other words, the duty not to be closed-minded or biased regarding novel scientific theories helps to increase the probability that novel theories will be recognized. This recognition, in turn, obviously affects the welfare of both proponents and opponents of the novel theories. 3) Finally, recognition of the duty not to be closed-minded in interpreting research results also presupposes an implicit notion of procedural justice. That is, it presupposes a correct method for arriving at any distribution of goods, a method that involves rules for awarding or withholding desired goods in a just way. Being biased for or against novel research results would interfere with procedural justice and with rules of just distribution because the bias could result in the least meritorious scientists receiving the best distribution of goods. (In the case of pure science, the relevant distribution of goods includes things like research monies and nonmonetary, professional support from one's peers.) If recognition of this duty not to be biased thus involves assent to a particular notion of procedural justice, then it must also involve a judgment about ethical values.

The *ethical* value, the duty not to be biased toward innovative theories, also contributes to science as a purely *epistemic* activity. This is clear if one considers what would happen were there no duty to avoid closed-mindedness; science would likely not progress as rapidly as it might, dogmatism and ideology might take the place of empiricism and testing, and erroneous doctrines such as Lysenkoism might become more the rule than the exception.

J. S. Haldane tells an interesting tale of what happened in one case of failure to be open to innovative results. The referees of the Royal Society rejected some groundbreaking papers of the nineteenth-century English chemist, J. J. Waterson. One of the referee's comments was that "the paper is nothing but of nonsense." Haldane wrote:

> It is probable that, in the long and honorable history of the Royal Society, no mistake more disastrous in its actual consequences for the progress of science and the reputation of British science than the rejection of Waterson's papers was ever made. The papers were foundation stones of a new branch of scientific knowledge, molecular physics, as Waterson called it, or physical chemistry and thermodynamics as it is now called. There is every reason for believing that, had the papers been published, physical chemistry and thermodynamics would have developed mainly in this country [that is, England], and along much simpler, more correct, and more intelligible lines than those of their actual development.[2]

As the remarks of Haldane illustrate, openness to innovative theories is clearly an epistemic value. But if the duty not to be closed to novel scientific results is both an epistemic as well as an ethical value, as was argued previously, then pure science may involve ethical as well as epistemic values.

Note that what I defend as the duty not to be closed to novel scientific results is similar to what Clifford discusses as the duty not to hold beliefs dogmatically.[3] My claim, however, is more defensible than, and different from, what Clifford calls the duty to avoid believing anything upon insufficient evidence.[4] Note also that I am defending a negative ethics of belief, a negative maxim about avoiding closed-mindedness, rather than a positive ethics of belief, a positive maxim, for example, about seeking the truth.[5] The former is more defensible.

OBJECTIONS TO CLAIMS ABOUT ETHICAL JUDGMENTS IN PURE SCIENCE

Although there are several objections to this negative ethics of belief, as I have summarized it, one of the most prominent criticisms concerns its presupposition that belief is under voluntary control, that one can voluntarily choose not to be closed-minded or dogmatic. Only if belief is under voluntary, human control does it make sense to say that purely cognitive decision-making is subject to moral appraisal. Because there are strong grounds for arguing that belief is not voluntary, and that failure to believe something is often not culpable, therefore (goes the objection) there are strong grounds for claiming that belief is not subject to moral appraisal.

In a recent article, William Alston gives one of the best defenses of the position articulated by this objection. Alston claims that "there are strong reasons for doubting that belief is usually, or perhaps ever, under direct voluntary control.... When I see a car coming down the street I am not capable of believing or disbelieving this at will."[6]

The problem with Alston's objection, however, is that most instances of belief acquisition, especially in research science, are not analogous to the case of seeing a car coming down the street. On the contrary, research situations are typically empirically underdetermined, at least in some respects, or else they would not be cases of genuine research—instances of breaking new ground, or instances of extending the limits of our knowledge. Because of this empirical underdetermination, and because scientific knowledge (especially at the frontiers, where the most controversial research is done) is always incomplete, all scientific results are subject to some degree of interpretation, at least in some small respects. But if so, then there are no uninterpreted scientific facts. There are no brute facts, and there is no fact/value distinction.[7]

But if all facts are interpreted facts, then believing them is, at least in part, a matter of choice. This is why earlier in the century, for example, eminently reasonable scientists, at least for some years, could be divided on the question of whether the data supported Pauli's postulation of the neutrino. This is also why, several decades ago, eminently reasonable scientists, at least for some months, could be divided on the question of whether cosmic ray data supported the postulation of the magnetic monopole.

Indeed, because of measurement and observability problems, especially at the level of very small or very distant phenomena (for example, those studied in high-energy physics and in radio astronomy), many epistemologists argue that one's beliefs about particular scientific data are essentially matters of convention, not empirical confirmation. New research results in high-energy physics, for example, are often interpreted primarily in the light of existing laws and symmetry principles, largely because direct empirical corroboration is difficult. Moreover, if someone doing high-energy research suddenly discovered phenomena that apparently contradicted an important symmetry principle, for example, scientists would not immediately give up the principle in question; instead they would treat the alleged falsification of the principle as an anomaly to be explained by some hidden variable or by some further theoretical refinement.

In other words, scientific principles and conventions do not necessarily take a back seat to empirical corroboration when scientists are interpreting research results. And if not, then particular beliefs about the results are in part a matter of choice and therefore in part a matter of voluntary control. And if so, then cognitive decision-making may be subject to moral appraisal, at least to the degree that it is subject to voluntary control. This reasoning suggests that everyday pure science, not just controversial research, like that of Waterson or Pauli, involves at least some instances of judgment and hence voluntary control.

The general reason why all pure research involves some degree of judgment or voluntary control, however, is that a values-based epistemology is necessary to explain our scientific inductive practices. In other words, our epistemic values (like external consistency or empirical corroboration) "determine what we ought rationally to believe," as opposed to what it is in our interests, or the interests of society, to believe.[8] To get a clearer notion of the role played by epistemic values in research, consider phenomena of pure science that are susceptible to different or conflicting interpretations. If there are two or more interpretations of the same scientific *facts*, for example, then researchers typically adjudicate this conflict by an appeal to a higher level of interpretation, that of *methods*. That is, they justify a given interpretation of the facts by appeal to a given method that is justifiably superior to another method yielding a different interpretation of the facts. Likewise, scientists typically evaluate different methodological judgments by means of appeal to the next higher level of interpretation, that of epistemic or cognitive *values*. That is, they assess conflicting methods by means of cognitive values like predictive power or explanatory adequacy or external consistency. They choose the method yielding the greatest predictive power, explanatory adequacy, and so on.[9]

The big problem arises when cognitive or epistemic values conflict, as in the case of Pauli. Pauli subscribed to the cognitive value of external consistency (with conservation principles) and therefore postulated the neutrino, whereas his opponents subscribed to the value of empirical confirmability/corroboration and therefore denied the existence of the neutrino. How do scientists decide which cognitive value is superior? Ellis claims that "the value of [external] consistency always overrides,"[8] but many empirically minded scientists would disagree and give the primacy to empirical corroboration. To decide which epistemic/cognitive values have primacy, it is clear that one must make an appeal to another level of values, either ethical or pragmatic or methodological. Because one must appeal to this fourth level of values, to resolve interpretational conflicts at the level of cognitive/epistemic values, and because there is no higher level, no court of appeal above this fourth level, value judgments are unavoidable, even in pure science. Hence virtually all pure science involves some partially voluntary choices, and hence some decisions that are at least in part ethical.

To this response, Alston and other opponents of an ethics of scientific belief make a second objection: "insofar as something [like a particular interpretation of research results] is chosen voluntarily it is something other than a belief."[10] (This objection is similar to that of William James: "these feelings of our duty about either truth or error are in any case only expressions of our passional life," not expressions of belief.[11]) This objection also fails, however, because it settles the question of whether there can be an ethics of belief by fiat, by stipulation. On this account, whether pure science essentially involves voluntary choices, and therefore ethical values, is not an empirical question but a definitional one. Alston defines beliefs as the sort of things about which there is no voluntary choice. Hence this objection fails.

Yet another objection to this brief sketch of the ethics of scientific belief might be that scientists' use of ethical values is accidental, not essential, to the practice of science. According to the objector, ethical values are accidental to science because some scientists can avoid being narrow-minded, dogmatic, or closed to innovative research results. If some scientists can avoid such ethical values, goes the objection, then ethical values characterize some people who practice science, not science itself.

This objection likely arises from the long and important tradition of attempting to keep values, especially emotive or bias values, out of science, so as to safeguard the ideal of scientific objectivity. Despite the importance of this ideal, however, it is obvious that some values, for example, cognitive or epistemic ones, are essential to the practice of science. Moreover, in at least one sense, it also appears that making judgments about ethical values is essential to science. The reasoning is as follows: Science cannot be done without scientists. Yet scientists as scientists must continually make judgments about issues such as theory acceptance and weighting available evidence. But when scientists as scientists must make judgments about theory acceptance and weighting available evidence, they must make decisions about how open they ought to be to innovative theories and anomalous data.

But decisions about openness in these matters are in part ethical decisions, primarily because 1) they have consequences for the welfare of the proponents of the innovative theories; 2) they presuppose particular notions of fair play and equal opportunity in the "game" of science; and 3) they presuppose an implicit notion of procedural justice. Just because a scientist is not following a biased ethics in making judgments that have these three types of ethical consequences does not mean that she is not following an ethics in science. In other words, even if a scientist is not biased or closed-minded, she is nevertheless making an ethical judgment if she makes a judgment that presupposes ethical notions or that results in consequences for the welfare of moral agents. Hence ethical judgments about openness to innovative scientific theories or data are essentially, although only in part, ethical judgments.

William James formulated yet another objection to this sketch of an ethics of belief; he claimed that it was idle to affirm that one has an ethical duty not to be dogmatic or closed-minded because no one knows when one really is being narrow-minded.[12] This objection is not wholly reasonable, however, and for several reasons.

For one thing, people often do know when they are being narrow-minded, both because of past instances in which they might have behaved this way and were later shown to be wrong, and because of disagreement or controversy that might surround the position they hold.

Second, although it is impossible, in general, to know when one is being dogmatic, it is often possible, in specific cases, to outline when one is adopting too narrow-minded a stance on particular assumptions or on interpretations that are empirically underdetermined in a certain way.

Third, although we are never morally responsible for what we are not aware of, nevertheless, it is obvious that we are morally responsible for our character development and for the habits we allow ourselves to take on. Hence, to the degree that being dogmatic or narrow-minded regarding research findings is the result of flaws in our character development, for which we are culpable, we are therefore responsible for our narrow-mindedness. This situation is analogous to that of a drunken driver who kills a person with his automobile and who then claims that he, having been drunk, was unaware of having hit anyone. Although drunks are not responsible for what they do not know, they are still responsible for the actions that lead to the situations in which they lose awareness. The drunk driver is responsible for taking the third or fourth or fifth drink, just as the scientist is responsible for failing to heed the advice of others, throughout her career, who suggested that she was narrow-minded.

Finally, it is also obviously wrong to say that one cannot be morally responsible for not being open-minded because, if one cannot be so responsible, then one could not be said to be culpable for the worst sorts of extreme biases. Moreover, if one were never said to be responsible for narrow-mindedness, then some dangerous consequences would follow: a person might not attend to his behavior, believing that it was outside his voluntary control, and hence the narrow-mindedness might increase. As a result, one might come to have less and less rational control over his behavior. But if so, then there are practical, as well as logical and epistemic, reasons for believing that one has a duty not to be narrow-minded in interpreting novel scientific results.

PRACTICAL CONSEQUENCES OF ETHICAL JUDGMENTS IN PURE SCIENCE

What consequences follow if even pure science is inherently involved with ethics? If these arguments have been correct, then even decisions in pure science have consequences for the welfare of others (at the least, for the welfare of scientists holding opposed views). Moreover, to the degree that research decisions made in pure science are probabilistically associated with particular applications of that science, then there is at least a probabilistic connection between pure science and the welfare of many humans affected by the application of that science. This is because, for any pure science, there is always a known, finite (albeit perhaps small) probability that the research will be applied in a particular way that affects human interests. The stronger the known probabilistic connections (between pure research and particular applications), the stronger the responsibility of the pure scientist for the consequences of the application of her research. But if so, then even the scientist doing pure research has an ethical duty not to be narrow-minded or dogmatic in doing her research; she has a duty to assess the probability that her work will be applied in an ethically questionable way.

It does not follow from this duty of assessment, of course, that given a high probability of negative consequences, the scientist ought not perform the pure research in question. This further inference, about not performing the research, would require a significant amount of ethical analysis. This analysis would have to be directed, in part, at determining the consequences of numerous other options and establishing the truth of many claims that are likely situation specific. Nevertheless, it is clear that the pure scientist cannot absolve herself from all ethical responsibility, whether for her science as pure science or for her science as applied. Just as there are no interpretation-free facts, so also there is no ethics-free or purely scientific research. Even pure science is laden with *prima facie* ethical obligations and judgments.

REFERENCES

1. For a defense of this point, see MCMULLIN, E. 1983. Values in science. *In* PSA 1982. P. Asquith, Ed. Vol. 2. Sect. 2. Philosophy of Science Association. East Lansing, MI.
2. RESCHER, N. 1980. The ethical dimension of scientific research. *In* Introductory Readings in the Philosophy of Science. E. D. Klemke, R. Hollinger & A. Kline, Eds.: 342. Prometheus Books. Buffalo, NY.

3. CLIFFORD, W. K. 1886. Lectures and Essays: 342. Macmillan. London; MICHALOS, A. 1978. Foundations of Decisionmaking: 209. Canadian Library of Philosophy. Ottawa.
4. CLIFFORD, W. K. 1886. Lectures and Essays: 346; MICHALOS, A. 1978. Foundations of Decisionmaking: 207ff. (note especially the discussion of the flaws in the Clifford claim that there is a duty not to believe anything upon insufficient evidence: 212).
5. For other authors who make this distinction, see MICHALOS, A. 1978. Foundations of Decisionmaking: 206; JAMES, W. 1956. The Will to Believe and Other Essays in Popular Philosophy: 17. Dover. New York, NY.
6. ALSTON, W. 1986. Concepts of epistemic justification. In Empirical Knowledge. P. Moser, Ed.: 30. Rowman & Littlefield. Totowa, NJ.
7. KUHN, T. 1969. The Structure of Scientific Revolutions. University of Chicago Press. Chicago, IL; HANSON, N. 1958. Patterns of Discovery. Cambridge University Press. Cambridge; QUINE, W. 1953. From a Logical Point of View: 40–53. Harvard University Press. Cambridge, MA.
8. ELLIS, B. 1988. Solving the problem of induction using a values-based epistemology. Br. J. Philos. Sci. **39**: 141–160.
9. For a discussion about resolving factual, methodological, and cognitive interpretations and disputes, see LAUDAN, L. 1984. Science and Values: 47–49. University of California Press. Berkeley and Los Angeles, CA.
10. ALSTON, W. 1986. In Empirical Knowledge: 31.
11. JAMES, W. 1956. Will to Believe: 17ff.
12. JAMES, W. 1956. Will to Believe: 30.

Remarks on Ethics and Values Studies at the National Science Foundation

RACHELLE D. HOLLANDER[a]

Ethics and Values Studies
U.S. National Science Foundation
Washington, DC 20550

Ethics and Values Studies (EVS) is part of the Studies in Science, Technology, and Society (SSTS) program at the National Science Foundation. SSTS results from a recent reorganization at the Foundation, placing History and Philosophy of Science and Technology (HPST) together with EVS. This aggregation and name change indicates the commitment of EVS and HPST to encouraging research in this area. It provides explicit recognition of the importance of this area, which is called by various names, depending on orientation, ranging from science studies, to science policy studies, to science, technology, and society studies.

Research in SSTS examines the structures and processes that govern the development and use of science and technology in their social, economic, and political contexts. EVS research and educational efforts focus on ethical or value aspects of the interaction between science, technology, and society. The purpose is to encourage thoughtful and systematic inquiry into the mutual influences of science or engineering and the moral life of individuals, groups, institutions, communities, and nations. EVS supports projects that enhance understanding of the social values and mutual obligations and responsibilities that arise in these interactions.

EVS-supported research projects often examine aspects of scientific or professional ethics; controversies surrounding effects of or influences on sciences and technologies; or ethical and value issues in developing and using scientific or technical tools for decision making. Cross-cultural research and interdisciplinary work are encouraged.

Projects exclusively focused on ethical issues associated with new military technologies and national defense strategies, or requiring the use of classified materials, are not eligible for consideration. This does not mean that all research of the kind discussed in this volume is ineligible, however.

I point with special pride to the contributors whose work has been supported through EVS; there are two in particular whose projects relate to this volume's topic. The work of one of the editors, Carl Mitcham, has been sup-

---

[a] Dr. Hollander is Program Director, Ethics and Values Studies in Science, Technology, and Society at the National Science Foundation. Her remarks represent her views, not those of the Foundation.

ported through EVS. Douglas MacLean, Paul Durbin, Kristin Shrader-Frechette, Deborah Johnson, and Carol Ann Smith have also received awards. With EVS support, Rosemary Chalk and Vivian Weil have directed projects on issues related to those being discussed in this volume, and I believe that Harold Relyea was associated with one or both of them. Alex Roland is a member of the HPST Advisory Panel.

Dr. Weil's project was titled "Ethical Implications of Trade Secrecy, Patents, and Related Property Controls for Science and Technology." The project examined arrangements that restrict or encourage the exchange of scientific and technological information. A conference brought together researchers, policymakers, and sponsors of research, to consider the ethical implications of these arrangements, their effect on research productivity, and their success in meeting their objectives. An edited collection of papers presented at the conference and other essays is being published. Another outcome from this: Dr. Weil has a related grant on scientific communication and First Amendment issues, from a private foundation.

The project Rosemary Chalk directed, under the auspices of the Office of Scientific Freedom and Responsibility (OSFR) at the American Association for the Advancement of Science, was titled "Openness and Secrecy in Scientific and Technical Communication." It explored the effects of institutional policies and professional behavior on the conduct of scientific and technical research. Particular attention was given to restricted communication practices resulting from national security interests, commercial concerns, and professional competition. The results of the project were reported in a special issue of *Science, Technology, and Human Values*, and were summarized in the final report sent to the National Science Foundation:

> The project found very little documentation to suggest that formal restrictions on scientific communication in the United States are growing at a rapid rate. However, the project participants consistently expressed concern that the potential for increased controls remains high. Participants indicated that there are much greater economic incentives and political pressures for researchers to withhold significant research information and materials in some areas of science. As a result, there is considerable suspicion that individuals or groups of scientists or engineers are more likely to engage in or to accept restricted communication behavior as standard practice in some fields of work previously characterized by openness.
>
> Informal controls, i.e. self-imposed restraints initiated by researchers in response to government actions or commercial or professional incentives, may be a more significant problem for scientific communication than the above-noted concerns. Most project participants agreed that the existing data base for evaluating informal communication practices is extremely weak, and that there are few indicators to monitor such behavior.[1]

This project provides an example of where and how the topic of this volume intersects with priorities for consideration in EVS. It also indicates a direction for future research.

EVS does not consider projects exclusively focused on military research ethics. The Foundation's primary responsibilities are to foster the health of civilian science and engineering and to promote science and engineering education. Because we do not support biomedical research, a similar stricture applies in that realm. Where ethical or value issues relevant to the Foundation's mission arise in association with military or biomedical research, however, it is appropriate for us to consider and, indeed, to encourage proposals. Coming to understand the nature of the value conflicts or synergies that exist or are possible in these overlapping zones is an important and appropriate part of the overall EVS task.

Right now, the OSFR is finishing a planning project that attempts to address the problem identified by the participants in their previous project—that we lack empirical information that would allow informal communication practices to be evaluated with respect to their effects on such value concerns as national security and economic interests, commercial concerns, professional competition and cooperation, scientific and technological progress, quality of the environment for research and education, and social and societal perceptions of science and engineering.

The planning project examined the potential for systematic research on how national security controls influence the conduct of research and the dissemination of unclassified scientific and technical information. The objectives of the proposed assessment would be 1) to determine the extent to which scientists and engineers understand the restrictions; 2) to assess how scientists and engineers attempt to balance the values of openness with protection of the national interest; and 3) to document the effects of such restrictions on research and the dissemination of scientific and technical information. In this planning phase, a team of consultants with expertise about the substantive and methodological issues, and about the perspectives of concerned constituencies, has been assisting the project staff—Mark Frankel and Deborah Runkle—in addressing the conceptual and technical difficulties associated with undertaking such an assessment. Staff and consultants have met to define and refine the research questions and to design and evaluate the research instruments and plans for data analysis. In the next several months, EVS expects to receive a report that will identify key research questions, instruments, and analysis plans, as well as uses that the assessment can serve.

The rest of this paper raises two matters for consideration. The first concerns the general area of the relationships between normative and empirical research. The second identifies some topics on which EVS could consider and encourage further research and, perhaps, education projects.

One of the difficulties and the strengths of EVS at the Foundation results from its stress on linking normative and empirical research. It seems to me that the meaning of normative terms is unclear without empirical work, that the meaning is part of the attitudes, behaviors, habits, and relationships that we see, or do not see. Perhaps this belief is a version of the statement that I know what I mean when I hear or see what I say. Many normative statements are intended to be descriptive as well as prescriptive. They are intended to describe morally appropriate phenomena. Thus, normatively or morally appropriate use of value-laden terms may depend on developing our empirical understanding. Furthermore, we cannot know what or whether normative prescriptions are necessary, or whether important normative problems or conflicts are in need of discussion or adjudication, or what kinds of problems there are, unless we have empirical information. This kind of information also enables us to judge the salience and importance of moral debates, and is essential if we want to develop solutions to important moral problems.

What are the implications of these last remarks for the development and submission of proposals to EVS? 1) Proposals that develop a line of research or build on prior work with both empirical and normative components are likely to be favored. Research following from such proposals often requires collaboration among researchers from different disciplines. 2) Formal and informal ways of involving persons representative of groups who should be informed about the results of the projects are important, to build credibility, to develop outreach, to devise language that different stakeholders can understand. Meeting this demand often requires investigators to develop contacts outside their own discipline and their usual communication channels. I do not underestimate the difficulties associated with these efforts. But it can be done. In part, the projects I have described, as well as this volume, present examples. 3) On the other hand, and very importantly, philosophers, historians, and social science researchers doing studies of science, technology, and society can identify, within these large-scale and problematic arenas, researchable questions that can be approached in their traditional manner. By locating their research within this larger framework, they too can make a strong case for support.

What issues are those in which EVS should exhibit an interest? I have indicated some of the criteria by which they can be assessed, and have provided some examples above. The examples concern issues that arise in the interactions of the patrons of research—government, industry, and foundations—with each other and with universities, individual researchers, and the research and teaching environments. How these issues get recognized, adjudicated, and resolved affects the moral behaviors of these individuals and organizations, and of course the larger community.

Military and civilian agencies and organizations—public and private and quasi-private—are research patrons. How their values influence the development and use of scientific and engineering research, how they interfere with or reinforce the values and goals of each of the actors in this process, as well as the broader social and societal values and goals we often share and sometimes disagree about, are researchable questions in EVS.

Some intersecting areas in which EVS research might be useful: management of radioactive wastes, computer science, robotics, process engineering, and (more generally) management of hazards. For instance, it would be interesting to trace the intersection of research on earthquake hazards and on verification of atomic tests. What rationales do researchers in these areas use to persuade their sponsors, and vice versa? What societal, organizational, and personal values do these positions reflect? What values do the positions promote or demote? Where do these values come into conflict? What are the economic, social, and moral costs and benefits of these emphases? Are there conflicts or synergies between military and civilian goals? What exacerbates or ameliorates these conflicts? Can we mitigate conflicts or promote synergies? Should we try? What moral problems do graduate students in these areas of science and engineering face? If we know that graduate students in certain areas of science and engineering are likely to face moral problems in making research and employment choices, can or should we help them make better choices? How can or should we help them and their faculties and administrations identify and manage those problems? How can we help scientists and engineers help their employers to overcome problems and promote ethical behavior in those organizations? How can we identify and promote better use of the results of such scientific and engineering research in responding to human and environmental needs?

No one project can answer many of these questions. Small and large efforts are needed. First, we need to identify what is known. Second, we need to specify what we do not know but could find out if we tried. Third, after demarcating the overlap between what we could know and what we feel we need to know, we may begin answering some of the questions just raised. This volume can provide an important stimulus to our thoughtful and creative efforts.

REFERENCE

1. CHALK, R. 1985. Secrecy in university-based research: Who controls? Who tells? Sci. Technol. Hum. Val. **10**: 3-119.

PART III. SPECIAL ISSUES AND DEBATES

The Just War Ethic and the Military Support of Science and Technology

J. BRYAN HEHIR

U.S. Catholic Conference
Washington, DC 20005

INTRODUCTION

My theme is the relationship between the just war ethic and military research, the latter of which, for the purposes of this paper, should be understood as scientific or technological research for or supported by the military. The basic point to make is that although the just war ethic has a contribution to make to a larger framework of thinking about the ethics of military research, it is, by itself, inadequate to respond to the question at hand. The just war ethic inevitably has to reach out to other ethical theories and perspectives and then, more importantly, must be supplemented by social-political analysis of the role of science in society, particularly in contemporary society.

Because the just war ethic may contribute to a moral evaluation of military research, but by itself fails to address the question at hand adequately, I will proceed by considering four issues. First, I will describe the rationale behind the just war ethic. Second, I will describe the relationship between the just war ethic and military research. Third, in light of the just war ethic, I will consider personal roles in institutional relationships. Fourth, I will argue that the just war ethic must be supplemented, among other things, by a theory of the professions as much adapted to the contemporary shape of American society as to military research.

RATIONALE OF THE JUST WAR ETHIC

Just war theory is sufficiently well known that reviewing details or repeating arguments from classic texts is unnecessary here. What is needed instead is a description of the guiding rationale behind this theory, in order to see how this rationale may relate to the more specific questions at hand.

In a strict sense, the just war ethic might be thought of as simply an ethical argument that tries to define when and under what circumstances the use of force is permissible. This is the way many people understand the theory. But the just war ethic actually has a larger meaning and purpose, as has been well argued by Paul Ramsey and James Johnson. If you see the just war ethic only in terms of a case-specific analysis *for* the use of force, you miss its wider rationale. The wider rationale or goal of the just war ethic is as an ethic of state practice. That is, the just war ethic tries to locate the use of force within a larger sense of the political process. Then it seeks to analyze the use of force both in terms of the purposes of the political process and within a framework of moral reflection.

To say that the just war ethic is an ethic of state practice is to emphasize that it has multiple moral purposes and that it is designed to limit and contain the use of force. The just war ethic is not *for* the use of force but *against* it. Its basic premise, in light of the wider sense of the political process, is that force is seldom permissible; that is, force may only be used in a few special circumstances. The just war ethic is not meant to justify but rather to limit and contain the use of force.

In order to do this, the just war ethic ties the justification for the use of force to a larger ethic, the ethic of the state. That is to say, it is central to any legitimate use of force by a state that the state be recognized as having political legitimacy, both internally by its own citizens and externally by the international community. Thus, according to the just war ethic, force may only be justified if the legitimate state or political authority that seeks to use force explicitly justifies its actions to the political community—not only to its own citizens, but to other states. This is the meaning behind the requirement that there must be a declaration of purpose about why force is being used. The declaration can then be examined both internally by citizens of the state and externally by the international community.

Finally, the just war ethic argues that even when a state can justify the use of force, it must go further and justify the means used to pursue some legitimate objective. The just war ethic is both an ethic of *ends* or the purposes for which force might be used and an ethic of *means*.

What one finds, therefore, is that the just war ethic argues that states have legitimacy, but that their legitimacy is tied to the way they conduct themselves within the political community. States are legitimate in principle, and have the right to use force under stringently limited circumstances, but they must justify the ends, and the means, to both their citizens and to the international community.

The argument that the state must justify the use of force to its own citizens has some relevance to the issue of military research. The just war ethic argues that the state must answer to not only a policy evaluation in its use of force, but the conscience of each individual citizen. The state may claim that it has a justifiable purpose in preparing to use force or using force, but individual citizens may on their own decide that in fact the state is mistaken or misguided, or that it is misusing force. There is thus contained within the just war ethic the possibility of dissolving the bonds of conscience between the citizen and political authority. In its full scope, the just war ethic is both an ethic of state practice and an ethic of personal choice. The latter deserves to be highlighted because in the contemporary world it has taken on particular relevance.

In Central America in the early 1980s, the Archbishop of San Salvador was assassinated after he explicitly asked the military forces of the country to disobey orders when they were asked to use force against civilians. Similarly, today in South Africa, the Catholic bishops of that country have argued on more than one occasion that there is a legitimate basis for citizens to refuse military service in principle because of the goals of the government. These responses to state policy are not mere critiques; these responses show state policy being approached at a different level, the level of the personal conscience of the citizen. If one accepts that the state is not only capable of abusing its authority, but of exceeding its authority, and that there is some overriding principle of justice manifest in the personal conscience of the citizen, then whether the state as a whole accepts an ethical critique may not be the most important point. The most important point may be that the just war ethic has a kind of capacity to undercut the ethic of the state by simply dissolving the rationale of the state in the minds of its citizens.

Thus, the just war ethic is not only an ethic of the state, or of how the state relates to the wider political and international community. It is also an ethic of personal conscience, which is a second kind of restraint on the state.

APPLYING THE JUST WAR ETHIC IN THE MODERN WORLD

During the modern period there have been a series of social changes that have presented problems for the just war ethic. I think the just war ethic is still a useful way to think about the morality of war and peace for states and individuals, but one must be careful not to underestimate the moral challenges that the just war ethic now faces.

The Question of Ends

The first challenge concerns the question of ends. At one time, the just war ethic fitted securely into a much more integral sense of political community. In the medieval world, for example, where this ethic had a strong influence, there was a wide sense of political community—the Christian community of Europe. Someone like Thomas Aquinas would argue that you could use force only after it was clear to the political community as a whole that one of its members was acting like an outlaw, so that the rest of the community would be galvanized in the style of a police force, if you will, to discipline the outlaw. There was a very strong restraint on this force because you had to demonstrate that someone stood outside of all recognized community norms.

The rise of the nation-state between the fourteenth and seventeenth centuries, however, dissolved the medieval vision of a political community. No nation-state recognizes the right of the larger community to judge the nation-state's final determination that it has reached a crucial moment justifying the use of force. This is the implication of the modern doctrine of sovereignty. Today, ethics require the state to submit any use of force to a test of legitimacy by the wider community, but the bonds of this wider community are much looser than they would have been in an earlier period.

The Question of Means

There are also two challenges concerning the question of means. One is a political challenge, and one is a technological challenge. The political challenge is a product of the nineteenth century. The earlier ethical argument was that the use of force always had to be limited. One way to limit the use of force was to make a distinction, a clear distinction, between people whom one considered open to attack, and people whom one considered morally protected from direct attack. One relied, that is, on the standard distinction between combatants and noncombatants. The moral distinction still holds, in terms of its intrinsic ethical legitimacy, but with the rise in the nineteenth century of the Napoleonic notion that when a society goes to war, everyone goes to war, it becomes much more difficult to sustain. This political challenge has been complemented in the twentieth century by a technological challenge; that is, in this century, the technology of war has again made it increasingly difficult to distinguish between combatants and noncombatants.

In particular, there is the challenge of the nuclear age. Even in standard nonethical political analyses, nuclear weapons break the link between the use of force and the political purposes for which force is used. The justification for the use of force was always that it served some political values. But nuclear weapons raise the possibility that they might, if used, destroy all values that could legitimate their use. This is the snapping of the Clausewitz connection. Clausewitz argued that war is the extension of politics by other means. Today, however, we face the challenge of the nuclear age. We must ask whether nuclear weapons could in fact ever serve the kind of political values that supposedly justify them.

This snapping of the political-military link in the use of force has, of course, implications for ethics. If it is difficult to justify the use of nuclear weapons politically, it is doubly difficult to justify their use in ethical terms. The ethical argument was that a limited use of force was the only legitimate use of force. The nuclear age raises the question of whether the limited use of force, if it includes the use of nuclear weapons, is not impossible.

The nuclear age raises another question, one that bears more directly on present concerns. In the past, ethical justification of the use of force became a real issue when war was entered. It was on the verge of war or in war that one raised questions about proper ends and proper means. The nuclear age, however, alters the moral issue. One must now justify not only war, but peace. In the nuclear age, to keep peace means relying on the threat of nuclear deterrence, which means one must be ready at all times to resort to a use force that almost everyone agrees could not be limited. Now, it is widely acknowledged that the same threat that we rely on to prevent war would be immoral if we were to actually carry it out. So, questions arise about the moral legitimacy of nuclear deterrence.

What do these questions have to do with the issue at hand? If we have reservations about nuclear deterrence, we might also have reservations about the preparations that sustain nuclear deterrence. To question the wisdom of carrying out these preparations, which involve the integration of science and technology into military strategy, is to confront the issue at hand—military research.

PERSONAL ROLES AND INSTITUTIONAL RELATIONS

If one has to think ethically not just about the use of weapons but also about their development—that is, about military research—what are the relevant institutions and who are the relevant individuals? How does one think about these in light of the rationale behind the just war ethic?

Government

The just war ethic always raises the question of the legitimacy of the state and its purposes and means. But nuclear strategy, to use it as a case study, raises the question of military research in a larger sense.

How does one think about this? One does not begin, it seems to me, by questioning the legitimacy of military research. One has to begin by questioning the legitimacy of the political project that stands behind the research, and the legitimacy of the means by which the political project of the state is to be implemented. I would argue that applying the just war ethic in this case serves mainly to raise questions about the wider purposes for which force is conceived or planned.

Arguing on the basis of the just war ethic raises two different possibilities. First, one could argue that all use of nuclear force lies beyond the pale of ethical legitimacy, and that therefore the strategy of deterrence in principle also lies beyond the pale of ethical legitimacy (according to this argument, any research necessary to promote the project is obviously called into question as well). Second, one could argue that the project of nuclear strategy and nuclear deterrence is marginally or just barely legitimate, and that one could not in principle rule out the research that sustains this project. One could, however, certainly raise a set of questions about specific aspects of the research. According to the Catholic bishops' pastoral letter on nuclear deterrence, for instance, it seems to me that the conclusion is that the strategy of deterrence is just barely legitimate, that is, not illegitimate. On this ground, one must at least acknowledge in principle the legitimacy of the supportive research, of the research that sustains the wider project that is just barely legitimate. At the same time, one is faced with a set of very stringent conditions for the means by which supportive research is carried out, and the ends for which it is intended.

Industry

To apply the just war ethic to an examination of the military–industrial relationship (that is, the military–industrial–scientific relationship), one must extend what was originally an ethic of statecraft and force, or an ethic of personal conscience and force, to a new area of society. Extending the argument used in the previous section to the military–industrial relationship, one may reason as follows: If one is not prepared to declare the whole project of the state illegitimate, if one allows that it is even marginally legitimate, then one cannot give up the supporting research or the supporting role played by the industrial sectors of society. Nevertheless, in the nuclear age it remains necessary to be sufficiently stringent in looking at the specific means by which the wider project is pursued, so that one does not translate "barely legitimate" into a free ride for a whole set of projects that are not really essential to it.

Universities

Universities are not the state and they are not industry. The university has its own particular integrity, its own particular purpose in society. The standards used to assess industry-based research, then, may not be appropriate to an ethic of university-based research. The university, because of its unique status, may require its own set of standards, which, perhaps, could encompass the standards for industry-based research. It does seem, however, that the standards for the university would have to be more stringent. In determining what kinds of research would be acceptable, all kinds of questions— for example, about classified research—could arise.

TOWARD AN ETHIC OF THE PROFESSIONS

When the just war ethic was formulated in the late Middle Ages, it was complemented by a fairly well developed ethic of the professions. It may be helpful today, in attempting to apply the just war ethic to the question of military research, to reintroduce the idea of there being an ethic of the professions. Doing so would bring us back to an idea mentioned earlier—the idea of there being an ethic of personal conscience, in addition to the ethics of the state, of industry, and of the university. Our idea of the ethic of personal conscience could, perhaps, be enriched by an examination of the roles people play in their personal lives, that is, by an ethic of the professions.

The professions were originally rather narrowly defined. Only three professions were recognized: law, medicine, and ministry. Today we have an explosion of professions and the corresponding necessity to expand the notion of the ethic of the professions to a whole new series of enterprises.

The core notion of the professions, however, is what can be called competence. Originally, the ethic of the professions was based on the idea that a professional could, by possessing certain unique skills, earn a certain status in society, this status being concomitant with a certain role in society. This role, in turn, was always tied to a notion of social responsibility.

A professional did not only pursue a particular discipline. He or she pursued a particular discipline in such a way that one understood that the pursuit of the discipline had an impact on society as a whole. So an ethic of research, military research, should account for the various professional disciplines that are encompassed by military research.

In the end, one must consider not only the legitimacy of state practice or the legitimacy of university participation. In the end, the just war ethic requires a return to personal questions, with these questions addressing the issue of professional responsibility.

CONCLUSION

Does the just war ethic by itself rule out, for example, scientific research for the military? Not unless one may say that the whole project of the state, politically and militarily, is unjustifiable. If one must allow that it is justifiable—if only marginally—then the just war ethic itself becomes marginal. The weapons that might be used today, and the political organizations with access to them, are such that the preparations necessary to create and maintain these weapons are just barely justifiable. The just war ethic here is being strained to the limit.

Although the just war ethic may be used to justify research, to justify research is not to justify everything. The just war ethic places stringent demands on what kinds of research are justifiable in light of wider political and military projects. Finally, an ethic that would be adequate today must be an ethic of the state, an ethic of other key institutions in society, and an ethic of personal and professional responsibility.

The Social/Professional Responsibility of Engineers

DEBORAH G. JOHNSON

Department of Science and Technology Studies
Rensselaer Polytechnic Institute
Troy, New York 12180

INTRODUCTION

One of the most difficult of the ethical issues associated with scientific research for the military has to do with a researcher's decision whether to become involved in such research at all, and if so, whether to draw a line between research she is willing to do and research she is not willing to do for the military. My aim in this paper is not to decide these questions, but rather to argue for a framework for thinking about them, a framework that would facilitate analysis and bring the important issues to the fore. The framework I propose places these matters in the context of broader questions about the social responsibilities of researchers.

The focus here will be on engineers though most, if not all, of what is said applies to other researchers. The special emphasis on engineers is useful for a number of reasons. First, social responsibilities are often understood to arise from membership in a profession because of the special expertise or unique role of the profession in society. Thus, focusing on a particular profession allows us to see how the very conception of a profession implies criteria for choosing projects. Second, a large portion of American engineers are employed by the military or by companies with contracts from the military, and a large portion of the research being done for the military is done by engineers. (There are important differences in the circumstances of academic engineers, corporate engineers, and government engineers, but these will not be emphasized here.)

From the ethics codes of engineering professional organizations and from the engineering ethics literature, it is clear that engineers generally see themselves as having responsibilities to society, to employers, to clients, and to coprofessionals or the profession as a whole. Many of the codes specify that the responsibility to protect the safety and welfare of society is paramount. It is this responsibility that we refer to when we talk about the social responsibilities of engineers. And it is this responsibility that comes into play in decisions about participation in military research.

This responsibility can be understood to fall on the shoulders of individual engineers and/or on the shoulders of the profession, that is, members of the profession collectively. In the latter case it falls to engineering professional organizations to take appropriate action (of a kind different from that taken by individuals) to protect the safety and welfare of society. Although the primary concern here is with individual responsibility, the role and importance of collective action should not be missed.

The question of this paper is, then, what do the social responsibilities of engineers call for when it comes to military research? The idea here is to derive, from the social responsibilities of engineers, a rule or principle that could help engineers decide what conditions should be met, if any, before they take part in military research. At first glance, it may seem that such a framework will not address the importance of individual conscience in the decision process, but it will become clear that this is not the case.

The paper aims to make some progress toward developing a principle that would apply to all engineers, and to do this by evaluating several principles that have been put forth in the engineering ethics literature. The three principles considered are roughly stated as follows:

1. Engineers should not impose their moral views on society; they should let society decide what projects are undertaken.
2. Engineers should refuse to work on projects they deem, on balance, to conflict with their moral values.
3. Engineers should refuse to work on projects that increase risk unless the public is informed about the risk and is given the opportunity to consent to the project.

THE GUNS FOR HIRE VIEW

The view suggested by the first principle can be called the "guns for hire" view of the social responsibilities of engineers because it suggests that engineers should make their skills available to whomever is willing to pay for them, presumably as long as the projects are within the law. Samuel Florman comes very close to adopting this view in "Moral Blueprints: On Regulating the Ethics of Engineers." In this article, Florman expresses concern about engineers imposing their own personal values on the rest of society, as well as society's willingness to rely on the authority of engineers so as to avoid having to make decisions:

> The problem-solver cannot factor his personal fancy into each equation. He must operate within constraints and expectations set by those who commission his work. . . .
> . . . Should we risk oil spills and increase our reserves by offshore drilling? Accept the hazards of pesticides in order to feed hungry people? Stop building a dam and thus protect an endangered fish? These are political questions: it is pathetic and a little frightening to see citizens abdicate their responsibilities by assigning them to the realm of engineering ethics.
> . . . If each person is entitled to medical care and legal representation, is it not equally important that each legitimate business entity, government agency, and citi-

zens' group should have access to expert engineering advice? If so, then it follows that engineers (within the limits of conscience) will sometimes labor on behalf of causes in which they do not believe.[1]

Florman's position sounds very much like Milton Friedman's view in "The Social Responsibility of Business Is to Increase Profits."[2] Friedman argues that the only social responsibility that a business has is to increase profits while staying within the limits of the law, and that corporate executives impose their personal views on stockholders and others when they use corporate profits to give to charity or to contribute to other social programs. Friedman claims that when corporate executives use corporate profits in this way, they, in effect, tax stockholders "without representation," and, of course, no one has the right to do this. Similarly, we might say of engineers that their only social responsibility is to refuse to work on illegal projects, and to do otherwise is to impose their personal values on the rest of the world.

On this view it is never wrong for an engineer to work on a project that has been endorsed by the public through the marketplace or filtered through government regulation. And, of course, this would seem to suggest that it is never wrong to work on a military project because military projects are presumably projects endorsed by society through representative government.

Perhaps Florman is right to worry about individual engineers imposing their personal values on society, but it is often impossible to separate personal from professional values and judgments. Professions are not value neutral, nor would we want them to be. The very idea of a profession implies commitment to a social good or value, for example, health, justice, truth, or human well-being. (Questions can be and have been raised about whether engineering is a profession, but I do not want to get sidetracked by that issue. At any rate, all sides of the controversy generally agree that the more engineering becomes a profession, the better.)

The problem with the guns for hire model is that it presupposes that individuals can and should become amoral in their professional roles. This contradicts the very idea of professions that are socially recognized and granted special privileges by the state, precisely because they are directed at a social good. We would not want a value-neutral profession of engineering any more than we would want value-neutral professions of medicine or law. We want doctors who are committed to a healthy society, lawyers who are committed to justice (as defined in our system), and engineers who are committed to safety and human well-being.

The guns for hire view of engineering is also objectionable because it assumes, as does Friedman, that the marketplace and government regulation will do a good job of filtering out projects that should not be undertaken. It assumes that anything that makes it through the filter is good for society. If this were so, of course, there would be no reason for engineers to have a responsibility for the safety and welfare of society. The "system" would, so to speak, bear the full responsibility. We know, however, that the system sometimes errs and that the marketplace and government regulation are not all-knowing and often lag well behind current knowledge relating to the safety and health of society. We also know that engineers (scientists, researchers, experts) are in a much better position to evaluate the safety and appropriateness of technological undertakings than are consumers and the public. They have the relevant knowledge and experience to evaluate the safety, credibility, and reliability of engineering projects. Of course, they are not all-knowing themselves. Their knowledge has limits. Nevertheless, it seems clear that we are better off if engineers act on our behalf, and not as guns for hire.

Thus, the guns for hire view of engineering is unacceptable. Engineers (and other researchers) should have a say about projects that are to be undertaken, and they should certainly not absolve themselves of responsibility for the effects of projects on which they work. It is in the interests of all that they conceive of themselves as having, and act as if they have, responsibility for the safety and welfare of society.

PERSONAL/PROFESSIONAL VALUES

In *Controlling Technology: Ethics and the Responsible Engineer*, Stephen Unger presents a model code of engineering ethics. Article 1 begins by stating that engineers should "regard their responsibility to society as paramount." One of the subarticles that follow states that engineers should

> endeavor to direct their professional skills toward conscientiously chosen ends they deem, on balance, to be of positive value to humanity; declining to use those skills for purposes they consider, on balance, to conflict with their moral values.[3]

Unger seems here to take a view in direct opposition to Florman's. The principle he asserts makes it clear that engineers should not be neutral. He asserts not only that engineers should refuse to work on certain projects, but that they should use their skills (perhaps only) for projects of positive value to humanity.

This principle is a significant improvement over Florman's account insofar as it commits all engineers to a shared value, the good of humanity. Of course, there will be disagreements about what is good for humanity; nevertheless, the importance of a commitment to such a value, however general the value may be, should not be underestimated. It defines the framework within which members of the profession must make their decisions. It provides the common ground upon which individual engineers or groups of engineers may disagree.

Still, while Unger asserts this value in the first part of subarticle 1.2, he seems to take it away in the second part, when he asserts that engineers should decline "to use their skills for purposes they consider, on balance, to conflict with their moral values." Here Unger seems to retreat from the idea of the profession having a value at its core, and to take the view that individuals are the only entities with moral values. In the discussion about the model code, he explains his concern that professional codes of ethics should leave room for individual variation and conscience, so that the profession does not constrain individual freedom. Still, the point is that there may be professional as well as personal reasons for declining to use your skills.

Although Unger's concern for individual freedom and individual variation within professions is appropriate, he draws an untenable line between personal and professional values and judgments. The principle says that one should not participate in projects that conflict with one's moral values, implying that one's moral values are personal and not professional. He separates concern for the good of humanity from one's personal moral values. When researchers refuse to work on projects or publicly oppose them on grounds of conscience (their own personal values), however, they generally do so because they believe the project is *not* good for humanity (or some portion of humanity). These are matters of professional judgment (at least partially) and not just matters of conscience. For example, those who have opposed the construction of nuclear power plants are surely expressing their technical judgment about the degree of safety obtainable in such plants. The Strategic Defense Initiative (SDI) is an even better example, for here the opponents have not been researchers opposed to all military research. Rather, they have been opposed on the grounds that the system could not be built to an appropriate level of reliability.[4] Matters of conscience and matters of safety and welfare of society are inextricably connected.

Because Unger does not recognize the difficulty of drawing a line between personal and professional values, he does not seem to recognize the significance of the profession itself being committed to a value. But, our concern here should not be with interpretations of Unger. What is at stake are two different conceptions of the place of professions in the world, and two different conceptions of engineering. One conception is to view professions as being value neutral, with the values directing them coming from outside. On this model it follows that engineers can and should be guns for hire. (Of course, on this view, there is little difference between professionals and nonprofessionals.)

The other conception is to view professions as enterprises committed to a social good and being structured (constrained in various ways, privileged in others) so as to achieve that social good. This conception maintains that some projects are antithetical to the profession itself and that participation in them would be "unprofessional." In this conception of a profession, and only in this conception, can we understand engineering as a morally worthy enterprise, worthy, that is, of individual commitment and social recognition.

Florman is clearly an adherent of the first conception. Unger is hard to place. He leans toward the latter, but does not seem to recognize the distinction.

Another problem with Unger's principle, related to the previous problem, has to do with its usefulness to engineers. Unger's principle is, to some extent, empty of content. That is, it does not tell engineers what their moral values should be. After all, the hard part about being moral is not just acting on one's conscience, but figuring out what one's conscience ought to be. It is precisely for assistance with this that a professional might turn to a professional code.

Unger does not seem to realize that he has begun to supply this content just by placing the matter in the context of the good of humanity. Of course, as mentioned before, there can be genuine disagreements about what is good for humanity, but this should not paralyze those who have expertise relevant to the good of humanity. In particular, it should not prevent them from making the good of humanity their end.

Thus, while Unger's principle is an improvement over Florman's insofar as it allows and expects individuals to remain moral beings in their professional roles, it is problematic for other reasons. It does not help engineers figure out what their moral values ought to be, and it does not recognize that professional values should play an important role here. It does not adequately recognize a value at the heart of the profession. In this respect, it pushes responsibility too much toward the personal realm and not enough toward the professional.

RISK AND PUBLIC CONSENT

In "Engineers Who Kill: Professional Ethics and the Paramountcy of Public Safety," a third principle is offered by Kipnis,[5] whose concerns are somewhat different from those of Florman and Unger. Kipnis argues that

> engineers shall not participate in projects that degrade ambient levels of public safety unless information concerning these degradations is made generally available.

Kipnis recognizes that risk pervades our lives and that we must accept this when we make decisions about engineering endeavors. He argues, in effect, that the real issue for engineers has to do with projects that increase the extant level of risk. And, he thinks the informed consent of those who will be put at increased risk by such engineering projects ought to be sought before such projects are undertaken. He does not, however, think that engineers, as engineers, have the responsibility for determining whether adequate consent has been given. Hence, he does not insist that engineers work only on projects for which consent has been obtained. Rather, he argues that "information about risks should not be withheld from the public" and that engineers should not participate in "covert degradation of public safety."

Kipnis's account is close to, and seemingly consistent with, Martin and Schinzinger's idea of engineering as social experimentation.[6] Martin and Schinzinger argue that engineering endeavors always involve risk. Engineers learn by trying something and seeing if it works, then learning from what goes wrong. We never know with certainty all the effects that will result from a project.

According to Martin and Schinzinger, once we recognize the experimental nature of engineering, we can model it on medical experimentation and draw on recent analyses of the ethical issues there. Analyses of medical experimentation make clear that researchers should not be allowed to experiment on subjects without their consent, that is, their informed and uncoerced consent. This imposes important responsibilities on medical researchers, and by analogy on engineers. Martin and Schinzinger conceive these responsibilities broadly. They seem to believe, however, that management and the public, as well as engineers, bear the responsibility to insure that those affected by a project are informed and given an opportunity to consent. Thus, while their account seems consistent with Kipnis's account, they shy away from asserting a principle of the kind Kipnis proposes.

The attraction of Kipnis's principle lies in its responsiveness both to Florman's concerns about individual engineers imposing their personal views on the rest of the world, and my concerns about the lack of content in Unger's principle. This third principle would not have engineers impose their views on the rest of the world. It would have those who will be affected by a project determine (or at least have a say in) what projects are undertaken. On the other hand, it does not make engineers into guns for hire because it imposes on engineers a distinct and substantive professional responsibility to insure that the public is informed about projects that will affect its well-being. Hence, it gives engineers professional grounds for refusing to work on certain projects. Engineers are in a good position to insure that the public is adequately informed because they understand better than nonengineers what the risks of a project will be.

The principle does not explicitly put forth the good of humanity as a value at the heart of engineering, but one could argue that it is implicit. It is implicit in the idea that those who are affected by a project should be informed and have a say. In effect, it specifies that whether a project is, on balance, good for humanity ought to be decided by humanity rather than engineers alone. This rightly suggests that the good at issue is not wholly a technical or engineering matter. It suggests that decisions about whether to take risks and decisions about how to balance risks against benefits involve more than technical expertise.

The problem with the Kipnis–Martin–Schinzinger approach is that the responsibility they assign to engineers is much weaker than their analyses would seem to indicate. This is, no doubt, because the ideas are so difficult to implement. Both analyses suggest that those who are being put at risk by a project ought to be informed about the risks and given the opportunity to approve or reject the project. But, this is wholly impractical, as Kipnis, Martin, and Schinzinger are well aware. How can we obtain the informed and uncoerced consent of the public? What do we do when there is less than unanimous consent (for example, when some people in a community think the benefits of a nuclear power plant or pesticides factory being located in their community outweigh the risks, and other people do not)? What do we do when there is some question about the voluntariness of the consent (for example, when there is high unemployment in the community in which many want the risky factory)? Will government approval count as representative consent? Will the political process be adequate for cases where there is a mixture of public approval and disapproval?

These are primarily practical, not theoretical issues, albeit very real ones. They lead both Kipnis and Martin and Schinzinger to propose something quite short of the full involvement of all individuals who will be affected by a project. It is important to note that this somewhat weakens the line between these authors and Florman—for if the political process, and perhaps the marketplace, were to count as indirect forms of consent, Kipnis, Martin, and Schinzinger would not be far from Florman. The public, through the government and the marketplace, would be deciding what projects were undertaken. Still, neither Kipnis nor Martin and Schinzinger seem to have quite this in mind, for they argue for broader responsibilities for engineers.

Thus, Kipnis's principle is still better than Florman's and Unger's principles, and closer to the mark in identifying and giving some substance to the social responsibilities of engineers. It would be even better, though, if it were explicitly coupled to a commitment to the good of humanity. Thus, another principle might combine the first part of Unger's principle (strengthened with the addition of the word "only") with Kipnis's principle roughly as follows:

> Engineers should only direct their professional skills to projects they deem, on balance, to be of positive value to humanity, and should refuse to work on projects that put people at risk, unless information about such projects is made generally available.

IMPLICATIONS

Now, what does all of this mean for an engineer's (or any researcher's) responsibility with regard to research for the military? The final principle indicates that a researcher contemplating participation in a military project should ask herself whether the project meets 1) the condition that the endeavor be good for humanity, and 2) the condition that those affected by the project be informed about it. Each of these questions brings an important issue of military research to the fore. The first condition raises the issue of how military research should be addressed. That is, should military research be assessed project by project, or should all military research be considered good (or bad) for humanity? The second condition raises the issue of how much secrecy in military research is tolerable. That is, can secret research ever meet the condition that information be publicly available?

Satisfactory analyses of these issues will require a good deal more analysis than can be undertaken here, though some important parameters seem clear. On the first issue, it seems very important to distinguish claims to the effect that research for the military, simply because it is for the military, is good or bad for humanity, from the claim that a particular project is good or bad for humanity. Recognition of this distinction is critical for intelligent discussion and coherent points of view.

The issue of secrecy is no less complicated. Some may argue that any principle ruling out secret research is unacceptable because there are times and cases when secrecy in research is necessary and, perhaps, even good for humanity. This might be so; nevertheless, a distinction might be drawn between the generic features of a project, and the details. With this distinction, we can insist that researchers should not work on projects the generic features of which have not been made public, though the details may well remain secret.

CONCLUSION

Both of these questions call for much more analysis and debate, but the aim here was to provide a framework for thinking about the issues associated with military research and not to settle them. The framework proposed is that of the social responsibilities of researchers. Although it is no simple matter to figure out what these are, it seems clear that researchers should not abdicate responsibility and act as if they were guns for hire. It also seems clear that we should not think that the ethical issues in military research are simply matters of personal conscience. They can be matters of professional judgment and professional commitment. The principle that I propose calls for researchers to consider whether their work is good for humanity and whether the public is adequately informed about the projects on which they work. These are considerations that ought to be at the core of the debate on military research. They ought to define the common ground, and thereby focus disagreement.

REFERENCES

1. FLORMAN, S. C. 1978. Moral blueprints: On regulating the ethics of engineers. Harper's. October: 30–33.
2. FRIEDMAN, M. 1970. The social responsibility of business is to increase profits. N.Y. Times Mag. September 13.
3. UNGER, S. H. 1982. Controlling Technology: Ethics and the Responsible Engineer. Holt, Rinehart & Winston. New York, NY.
4. PARNAS, D. L. 1987. SDI: A violation of professional responsibility. Abacus 4(2): 46–52.
5. KIPNIS, K. 1981. Engineers who kill: Professional ethics and the paramountcy of public safety. Bus. Prof. Ethics J. 1(1): 77–91.
6. MARTIN, M. & R. SCHINZINGER. 1988. Ethics in Engineering. 2nd edit. McGraw-Hill. New York, NY.

Conducting Scientific Research for the Military as a Civic Duty

KENNETH W. KEMP[a]

Department of Philosophy
Texas A&M University
College Station, Texas 77843

INTRODUCTION

In recent years, many scientists and other academics have treated scientific research for the military with suspicion. Although few would argue that no such research should ever be done, many are concerned that it (or at least much of it) should, at a minimum, be excluded from the university. (Tela C. Zasloff's article in the *Chronicle of Higher Education*[1] is a recent example of this.) In this paper, I want to argue that that is the wrong way to think about scientific research for the military. The scientist should instead see military research as a kind of civic duty. Although not all scientists will need to act on this civic duty, and there are circumstances under which doing such research would be wrong, some of the most common objections against military research carry less weight than is sometimes given to them.

SCIENTIFIC RESEARCH FOR THE MILITARY AS A CIVIC DUTY

I want to begin by making explicit two assumptions that, although they seem to me to be true, are certainly not uncontroversial. Discussing them would be of philosophical value, but doing so in any detail would keep us from getting to the central concern of this paper. So I will just assert them with only a hint of what would have to be said in their defense.

The first assumption is that war, or at least preparedness for war, is (both in general and in the present political situation) a legitimate means of pursuing just policy objectives. This means that both absolute and modern-war pacifism are false. A defense of the right to go to war would require defense

---

[a] Present address: Department of Philosophy, College of St. Thomas, 2115 Summit Avenue, St. Paul, Minnesota 55105.

of the view that the duty to see that justice is done (or injustice avoided) sometimes overrides the duty of nonmaleficence.

The second assumption is that it is the duty of the government to protect society from certain kinds of aggression, and it is a duty to do so by waging war if war is the only effective means of providing that protection. The duty of the government to wage war follows from a fairly specific right to be rescued. Although the question of whether there is a right to be rescued by passersby is perhaps controversial, the same right as a claim against the government is, I believe, easier to establish. For defense against the threat of force is precisely one of the reasons why a government is worth having. If there is a right to kill in the cause of justice, and a government duty to exercise this right (for example, by protecting its citizens from attack), there will, in certain cases, be a duty to go to war.

Given those two assumptions, there are three reasons for thinking that the scientist has a duty to participate in military research. The first reason, and the one perhaps most properly tied to a *civic* duty, is the government's just-mentioned right and duty to defend its citizens against aggression. Scientific research can improve the government's ability both to wage war, when war is necessary, and to deter war when war is threatened. Improving the government's ability to wage war makes it easier for the government to fulfill this obligation.

This claim about the right and duty to wage war, however, is not strictly necessary to my thesis. To the extent that a nation's military forces are primarily (or exclusively) deterrent (as perhaps are those of Switzerland and Sweden), and if the Wrong-Intentions Principle does not apply to deterrent threats (as I have argued that it does not[2]), it might be possible even for a pacifist scientist to support military research. But defense of the legitimacy of military research in the context of nonpacifist views about the moral permissibility of war is surely the more natural way to approach the issue.

But the government's duty to go to war cannot just be the duty of an institution. It must ultimately become the duty of particular individuals. Those individuals must be either mercenaries, a military caste, or ordinary citizens. Hiring mercenaries raises a host of moral and practical problems. Reliance on a military caste can endanger the principle of civil supremacy over the military. That leaves defense as the duty of ordinary citizens.

The days of the militia are gone, probably forever. There is no alternative but to maintain a permanent standing army. This creates the problem of assuring that the army maintains its allegiance to the ideal of the citizen-soldier, that it does not become a military caste. There are a variety of ways of doing this. For its part, the military can and must keep careful watch on itself, to ensure that officers with Bonapartist views are not promoted to positions of responsibility. If for no other reason, it has this duty because, in the United States at least, it sees itself, and asks others to see it, as a profession and not merely as an occupation. (See, for example, many of the essays in

M. M. Wakin's *War, Morality, and the Military Profession*,[3] which is used in the required ethics course at the U.S. Air Force Academy.) But the citizenry also has a part in this. Its responsibility is to ensure that civilians do not shunt the duty of national defense off onto the military as though defense were not a general responsibility. Encouraging those who have no intention of making a career of military service to spend a few years on active duty is one way of keeping the responsibility general. Broadening the involvement of civilian scientists in military research is another.

From the scientist's point of view, doing scientific research for the military will be just one way of fulfilling the rather stronger duty of all citizens to help make the society a better place to live. A scientist could fulfill this duty by being a soccer coach, by feeding the poor, or by raising funds for the symphony. But some scientists must fulfill this duty by doing scientific research when such research is needed. Indeed, because scientific expertise is so scarce, it may be important that many scientists fulfill their duty in this way. If enough scientists take their turn at such research, the burden on any particular scientist need not be too great. (Indeed, because military research has, in recent years, been at least as well funded as other scientific research, the burden of doing such research amounts only to the distraction from what might be intrinsically more attractive lines of research. In comparison to the burden imposed on the conscript, or even the volunteer enlistee, this burden is modest indeed.) Whether the scientist who decides to fulfill his duty to society as a scientist should fulfill it by doing scientific work for the military or by working on some other project will depend, in part, on what work most needs to be done. But the duty imposed by scientific expertise cannot be ignored; it is analogous to the lawyer's duty to do *pro bono* work for indigent clients.

There are two other reasons why scientists have a duty to assist in military research. First, when the nation asks some soldiers to risk their lives in pursuit of certain national security objectives, it has the obligation to minimize those risks by making available to the soldier whatever technological advantages will increase his prospects of survival. That imposes on the government, and consequently on society in general, a duty of research into chemical protective equipment, armor, night observation devices, and the like.

Second, the scientist has a general obligation to minimize the destructiveness of war. Military research aimed at this end will range from increasingly accurate weapons (which reduce collateral damage) to improvements in military medical technology (which the medic is required to use for the benefit of all wounded combatants).

Most of the objections to military research are focused not so much on the rejection of any of the premises of the foregoing argument as on other concerns, concerns which, it might be argued, outweigh those just advanced in favor of broadly based scientific participation in military research.

UNJUST CAUSES, UNJUST MEANS, AND MILITARY RESEARCH

The right and duty on which the argument in the preceding section was based was the right and duty to wage war, whether in self-defense or in defense of some other right. It goes almost without saying that scientists would have no duty to conduct military research in nations whose foreign policy was dominated by the pursuit of unjust ends. Indeed, ordinarily they would have no right to do so.

The traditional moral rule about cooperation in the wrongful actions of others distinguishes two kinds of cooperation, material and formal. Formal cooperation involves acceptance by the cooperator of the wrongful intentions of the principal agent. An example of such cooperation would be the actions of the conspirators in the Lincoln assassination. Even those who did not actually do the shooting were responsible for the assistance they provided to Booth because they were helping him precisely because they wanted the assassination to succeed. Formal cooperation in the wrongdoing of others is, of course, always wrong. Material cooperation, by contrast, involves cooperation in the *action* of the principal only. It does not involve acceptance by the cooperator of the wrongful intentions of the principal even when he knows of those wrongful intentions. An example would be the action of the mechanic who repaired a getaway car under duress. Such an action would be wrong unless 1) the action was not wrong in itself (as killing an innocent person would be but repairing the car would not) and 2) there was sufficient reason for performing the action despite the fact that it would help in the commission of a crime (as preserving one's life would be, but making a few extra dollars would not).

Scientists who performed military research for aggressive states would surely be doing wrong if they conducted the research in order to further the aims of their government. And although someone *might* be able to think up circumstances in which military research in the service of aggressive nations would be acceptable, on the whole, the *prima facie* duty to avoid even material cooperation in the wrongdoing of others would require avoidance of such activities as well.

The same would have to be said about research into intrinsically objectionable weapons (it would not, of course, be morally objectionable to do research into the means of defense against such weapons), as well as research in violation of treaty obligations.

MILITARY RESEARCH AND MISPLACED EXPENDITURES

Is there a moral obligation to refrain from engaging in military research that should have been put aside in favor of other societal needs? Is this also wrongful cooperation in the misdeeds of others? Surely, encouraging unnecessary military spending is blameworthy. For those in positions of influence, even failing to speak up against it may be blameworthy as well. But once the decision to undertake certain research has been made, there is ordinarily no reason why any scientist who believes that he can do the work well should not offer to do it. Here we have a case of merely material cooperation. It is probable that even those who proposed the expenditures did so because they were ignorant, or at worst negligent, of the injustice involved. So, in the ordinary case, there is little likelihood of the cooperating scientist adopting any of the wrongful *intentions* of the principals. Further, engaging in ordinary military research is not wrong in itself. And there are several reasons why the scientist should *not* avoid such material cooperation. First, once the money is allocated, the question becomes one of ensuring that the research is done as well and as economically as possible. Second, if all scientists who favor relatively lower defense expenditures avoid military research on the grounds that military spending is excessive, there will be a kind of concentration of military research in the hands of those who share certain political views. This would create precisely the kind of narrow political base of military work that should be avoided.

MILITARY RESEARCH AND THE PRINCIPLE OF POLITICAL NEUTRALITY OF UNIVERSITIES

Does participation in military research violate any kind of principle of political neutrality for universities? Would it, in consequence of some such principle, be inappropriate for university scientists to engage in such research? Again, I believe, the answer is no.

Just what a principle of political neutrality requires and forbids is hard to say. Some kinds of activities are obviously forbidden. Hiring faculty on the basis of their political beliefs, endorsing candidates for political office, and conducting political campaigns on behalf of particular candidates are all clearly violations of the principle. Taking sides on particular political issues, at least those far removed from education policy, is only somewhat less controversial. Even those prohibitions have some bite with respect to scientific research for the military, however. A university that conducts military research would, I think, be violating the principle if it waged a lobbying effort in favor of continued funding of that research. For whether such funding is necessary, and whether it is more important than other projects that are competing for the same scarce government money, is a political decision. (This restriction does not, of course, apply to individual academic scientists, who have a right and a duty as citizens to contribute to the debate over these issues. They would also, of course, have a special obligation to be careful lest their personal interest in seeing the funding continue cloud their judgment about its relative merits compared to the merits of other possible uses of government money.)

But in doing military research for a just state, the university only provides the government with the means that the government has chosen to develop for a more effective pursuit of its foreign policy. And although the choice, even among means, may be a political issue, the university could hardly be said to be *making* political choices in virtue of its agreement to help implement choices once they were made.

CLASSIFIED MILITARY RESEARCH AND THE PRINCIPLE OF OPENNESS

Some, though not all, military research is classified. Does conducting such research violate a general principle of openness that should characterize either all scientific research, or at least university-based research?

Surely the ideal of an open and unsecretive community of researchers who share results and ideas freely with one another is a noble ideal. The scientific community is at present not such a community, and in part, at least, for reasons that have nothing to do with national security or even with industrial secrecy. James D. Watson's *The Double Helix* offers a good account of the extent to which the eagerness for priority keeps the spirit of provisional secrecy alive in the scientific community. But perhaps what he describes just reflects the vanity of scientists, or the improper pressures imposed by the present institutionalization of science (where continued employment as a scientist requires publication).

Perhaps so, but there are nevertheless good reasons for keeping some scientific results secret. In an ideal world, there would be no knowledge that would be too dangerous to publish, but ours is not that world. In fact, there are things and processes the knowledge of which would enable bad men to pose serious dangers to society. A scientist who stumbles onto such knowledge would have an obligation to restrict access to it or even, under certain circumstances, to suppress it altogether. I think this obligation is analogous to the (unfortunately no less controversial) duty of a journalist not to publish certain kinds of information (for example, gratuitous dirt on public figures). Although there is knowledge that would be dangerous in the hands of some, it does not follow that such knowledge should never be sought at all. It might, after all, provide an important advantage to an honest police force or to an army defending its country. Thus there might remain an important duty to acquire the dangerous knowledge, but to make it known only to those who would not use it for the wrong purposes. If that is true, the only remaining problem is whether there is something distinctive about the university that would make it necessary that this kind of work be done elsewhere.

A recent University of Michigan report suggests that there is something special about universities:

> Research that is secret either in its conduct or its results is directly incompatible with fundamental University goals of the furtherance of knowledge through the discovery and interchange of scientific information and its faithful transmission to the younger generation in the classroom and laboratory.[4]

The report later cites several reasons why openness is important. Surely the most serious is that avoiding secret research "ensures that all members of the university community are free to participate in the research enterprise, including . . . foreign nationals." As long as scientists have the opportunity (both by university policy and in fact) to pursue classified research in laboratories outside the university, it is possible, as the Michigan Report suggests, that the separation of classified research and the university may turn out to be the best way of respecting all of the competing values that come into play here. But at this point the question is not whether the scientist will have an opportunity to fulfill his duty, but only of establishing an appropriate institutional setting for his work. (Margaret Schabas[5] has discussed the question of whose right it is to make decisions about the proper institutional setting for such research.) And that, unlike the question of the academic scientist's right or duty to conduct such research, is a question of institutional policy, not of moral principle.

CONCLUSION

Scientists must avoid the temptation to see military research either as a distraction to be avoided or as an inherently suspect category of work. The defense of the nation continues to require military preparedness, and just states have both the right and the duty to defend themselves. Derivatively, the citizens of those states have the duty to assist in this preparedness when and if their talents are needed. Although the duty to do this can be overridden, it is not overridden as readily as some recent critics of military research have suggested.

REFERENCES

1. ZASLOFF, T. C. 1988. The university, by definition, may be the wrong place for military research. Chron. Higher Educ. 34: A52.
2. KEMP, K. W. 1987. Nuclear deterrence and the morality of intentions. Monist 70: 276-297.
3. WAKIN, M. M. 1978, 1988. War, Morality, and the Military Profession. Westview Press. Boulder, CO.
4. *Ad Hoc* Committee on Classified Research at the University of Michigan. 1986. Report of the committee. July 14. University of Michigan. Ann Arbor, MI.
5. SCHABAS, M. 1988. The permissibility of classified research in university science. Public Affairs Q. 2: 47-64.

Military Funds, Moral Demands
Personal Responsibilities of the Individual Scientist

DOUGLAS P. LACKEY

Department of Philosophy
Baruch College and the Graduate Center
City University of New York
New York, New York 10010

MORAL CRITICISMS AND NONMORAL APPRAISALS

I want to begin this essay on a cautionary note. My topic is the ethical implications of the military funding of scientific research, but I want first to get clear in my own mind about what an ethical issue is. It has been my experience that what many people will call an ethical implication does not correspond to what I take to be an ethical implication, and arguments often go at cross-purposes.

I think that we should try to get clear about what ethics is because genuinely ethical judgments are both action-guiding and overriding. It is part of the logic of the terms "ethical" and "moral" that if something is unethical or immoral, it ought not to be done: the immorality of an action provides a conclusive reason for not doing it. The fact that moral judgments override all other judgments stems from the deep connection between what is moral and what is rational for human beings to do, given the social character of human life.[1] So it would be a mistake to think that a moral consideration is one of several considerations for or against a policy or action. On the contrary, a moral judgment is a final consideration about what ought to be done, or what one ought to do, all things considered.[2]

This view of moral judgments, quite common among philosophers, is accepted, for bad reasons, by people who still believe that the moral law is the law of God, but it is quite out of fashion among allegedly educated nonphilosophers. At meetings of the Institutional Review Board of a large metropolitan hospital on which I serve, the board will debate the merits of a research protocol involving human subjects and then turn to me for the "ethical considerations." I supply them, and the committee factors the "ethical considerations" into its decision about whether to sanction the research. I accept the bureaucratic logic of this arrangement, but it appalls my deeper philosophical sensibilities.

Moral judgments—again because of their rational character—are essentially universalizable.[3] What this means is that if a particular action is morally required of a certain person, it is morally required of every other person who stands in the same morally relevant circumstances as the first. A mere difference of persons counts for nothing as regards moral obligations. Two similarly situated persons may rationally adopt two different life-styles, but each cannot consider himself bound by different obligations than the other. Philosophers have made a great deal of the requirement of universalizability, but I am not so much interested in universalizable judgments being moral as I am in nonuniversalizable judgments not being moral; that is, nonuniversalizable judgments cannot express moral requirements.

Given these logical constraints on moral judgments, a distinction naturally emerges between an appraisal based on a political ideal and a moral criticism based on moral principles.[4] We might have certain ideals for ourselves, or others, or for our society, and we might praise those whose conduct serves these ideals. Nevertheless, it remains possible that conduct that falls short of these ideals is not necessarily immoral conduct. To take some examples, close at hand:

1. We might argue that it would be better if there were no such thing as the military funding of scientific research because such funding clashes with the desire of nonmilitary researchers to engage in the disinterested pursuit of truth. But the disinterested pursuit of truth is an ideal, not a moral requirement. Human beings are not obliged to disinterestedly pursue the truth. We may admire those who pursue truth, but we do not number among the wicked all who have no interest in doing so. Thus our appraisal here does not amount to moral criticism.
2. We might argue that the influence of the military establishment on the direction of scientific research in this country, considering the volume of scientific work devoted to military projects and the volume of scientific work sponsored out of the military budget, has had a debilitating effect on the ability of the United States to compete effectively in world industrial markets.[5] But even if we suppose that this is true, and I do suppose this, we must recognize that the economic supremacy of the United States is not something demanded by basic moral principles. American supremacy is not a moral good but a patriotic ideal. Once again, our appraisal is not a moral criticism.
3. We might feel that the scientists and students who signed the petition not to take funds for research budgeted through the Strategic Defense Initiative (SDI) performed a morally admirable action. But unless we are prepared to condemn all those who failed to sign the petition, including those who did not sign the petition but who are in fact *not* working with Defense Department funds, the action of those who signed the petition is not universalizable. We can say that those who signed the petition were pursuing a personal ideal, an ideal that we happen to find admirable, but we cannot say that they were discharging moral obligations when they signed.

WHAT IS IMMORAL IN THE MILITARY FUNDING OF RESEARCH

If we concede that the above-mentioned arguments (that the military funding of research is bad for science, bad for the American economy, and bad for the souls of idealistic young researchers) do not amount to a moral condemnation, we might think that the moral slate on military funding has been wiped clean. But in fact I do not think so: I think that the military funding of research is bad, morally bad. I begin my analysis from the standpoint of those who accept the funds.

First, it is obvious that those who accept these funds believe either that the research will in the end yield something that is militarily effective, or it will not. Second, those who believe that their research will in the end yield something effective believe either that this something will be used for good or it will be used for evil.

Case 1

Consider first those who believe that the research is militarily useless. This is a common enough view among those applying for SDI funds because most believe that the goals of the initiative simply cannot be achieved. What I want to say about these people is that they are engaged in a kind of theft: they are taking money from the taxpayers or their representatives on the pretense that they can produce what the taxpayers want for this money, knowing full well they cannot produce it. They are morally akin to meretricious doctors who supply deviant cancer treatments for pay, knowing full well that the treatments will not work.

There is a further error involved in the "Stars Wars won't work, therefore Star Wars research is benign" argument. Even if the goals of SDI cannot be achieved, it may happen that research toward these impossible goals can be diverted for evil purposes. The recent leaks that Star Wars technology will be directed to the destabilizing mission of antisatellite warfare is a case in point.[6]

Case 2

There are some scientists who are willing to concede that their research serves a military function and that this function is evil. In these circumstances there is no moral justification for doing the research: the need for scientific progress provides no argument because scientific research, as I have already urged, is not a moral imperative. Persons in this category usually feel that they have some *excuse*, if not a *justification*, for what they are doing. Whether they have an acceptable excuse will be taken up in the next section.

Case 3

Finally, there are those who believe that their research will serve some military purpose and that this purpose is benign, or perhaps even morally good. I do not wish to challenge the sincerity of those who hold these views, but I do challenge their accuracy. Although the question is daunting, we must ask and seriously try to decide whether the use of military force by the United States in the present era has been morally benign, and we must especially ask whether those aspects of military force that have been developed with the assistance of top-line science and engineering have been morally benign. We must try to approach this problem with at least a minimum of objectivity, defining that minimum as not taking the position that a policy is morally right by definition because it is the United States that has adopted it.

The historical record, at least the record since the Korean War, is not encouraging. In 1953, the American government acted to prevent the realization of legitimate nationalistic aspirations in Iran; likewise, in 1954, it overthrew the Guatemalan government on the pretext that any attempt to regulate the overseas operations of the United Fruit Company was a plot devised by Leninists. In 1956, the United States subverted the 1955 Geneva Agreements by supporting Diem's refusal to hold elections in Vietnam. In Cuba in 1961, in the Dominican Republic in 1965, in Chile in 1973, and in Nicaragua since 1980, the United States has supported forcible attempts to overthrow popular and populist Latin American governments.[7] The late 60s and early 70s were devoted to a foredoomed attempt to block Vietnamese self-determination, and since 1967 the United States has devoted massive resources to blocking Palestinian self-determination. In three of the most violent convulsions of the post-World War II era, the United States tilted to the unjust side: in 1971, in siding against India and the Bengali Independence movement; in 1978, in siding with Pol Pot and the Khymer Rouge against the Vietnamese; and since 1980, in siding with Iraq against Iran, although it was Iraq that invaded Iran, and Iraq that pursued this conflict in vicious disregard of the conventions of war. In all these cases I am prepared to argue that the threat or use of force by the United States was, putting it mildly, not benign.

I am not maintaining that the United States is the Great Satan, the prime offender against international justice. Certainly the record of the United States as regards the use of force since 1953 is not morally worse than that of the Soviet Union, nor is the use of force by the United States, measured by historical standards, uniquely or even especially malevolent. Nevertheless, the record is sufficiently depressing and consistent to make one think that the difficulty is not the result of bad judgment or accidental circumstances but of deeper structural factors, such as the maintenance of global political influence, the needs of the American economy, and the deep-seated myths the American people have about their moral position in history and in the world. There is also the general problem that the use of violence is only justified as a last resort, and politicians, American and non-American alike, rarely have the patience to exhaust all available resources before turning to violence. Hence, it is a safe historical generalization that most uses of violence by states are unjust. If the historical record since 1953 results from such structural factors, we have every reason to expect that uses of force by the United States in the near future will be as immoral as uses of force by the United States in the recent past.

If we turn now to those aspects of the use of American force in which scientists have been especially involved, we find once again structural factors that create a strong probability that scientific research will be used to ill effect. I do not wish to debate the pros and cons of particular weapons systems or to delve into the intricate immoralities of research on chemical, biological, and nuclear weapons. What concerns me is that the application of technology to warfare seems in the present historical phase to create an increasing distance between the killer and the killed, and this increasing separation between the killer and the killed must inevitably lead to bad judgments about the legitimacy of war and bad judgments about how to conduct it. The Vietnam example is the first case that comes to mind, and the last one that anyone should need on this subject. Furthermore, by and large, high-tech armaments in contemporary settings are more suitable to offense than defense, and low-tech armaments are more suitable to defense than offense. Because most offensive wars are immoral wars, research on the high-tech end is more likely to be used for evil than for good. Please note that my argument here is probabilistic, and is not to be overthrown by the example of British radar or the other standard stories of nice technology at work.

My verdict is that those who believe that their research will probably be used for good purposes are mistaken in this belief. If I am right, their participation is not morally justified. Once again, the question of excusability is withheld for the next section.

I have now exhausted the set of possible cases, and the result in the three main cases is that the acceptance of funds is not morally justifiable. But if it is not morally legitimate for anyone to accept the funds, it follows that it is not morally legitimate to offer them either. A policy that encourages immoral action by individuals must itself be an immoral policy.

THE POVERTY OF EXCUSES

This leaves the residual problem of whether scientists and engineers who participate in research under military auspices have some moral excuse for their participation.

First off, let me note that the traditional excuses supplied by the history of ethics—coercion and ignorance—are ill-suited to the typical case of military research or military funding.

The typical scientist cannot say that he is coerced into doing the research. It might be the case that he will be better off, in financial and career terms, if he does the research than if he does not do it. But the prospect of losing some gain that one might have had, as opposed to suffering some loss of what one already has, has never qualified as an excuse in either morals or law. Of course, there will be exceptions, such as the case of the company scientist who is transferred, on pain of dismissal, from a nonmilitary to a military project. This scientist, in my view, has an excuse, but his is not the typical case.

Likewise, the excuse of ignorance does not come readily to hand. In case 1 and case 2 above, where the scientists know that the research will be ineffective or pernicious, the excuse of ignorance is obviously unavailable. Case 3, in which the research is pernicious but is perceived to be benign, is more difficult, but I rush to point out that the mere belief that one is engaged in a benign activity does not, by itself, excuse the activity unless this belief is reasonable, that is, based on decent evidence decently examined.

It has not been my impression that many scientists involved in defense work or subsisting on the defense dole have undertaken an examination of the historical evidence and the present world scene. On the contrary, one finds, given the educational attainments of these people, a surprising ignorance of history, and an incredible lack of exposure to alternative interpretations of what happened and why. William Broad's account of the political naivete of young scientists at the Lawrence Livermore National Laboratory describes not the odd case, but the typical ignorance of history and the humanities one finds among these researchers.[8] I am not requesting that everyone slavishly accept the account of recent history that I have sketched here. But, given the impact of science on war, it should at least be possible for those who take defense money to present a modest point-by-point rebuttal of the cases I mentioned, and this they usually cannot do. What we have is culpable ignorance, not the nonculpable ignorance that generates acceptable moral excuses.

There are, however, more exotic excuses available to the scientists and technicians involved.

The Replaceability Excuse

One excuse, which I heard frequently from men of draft age during the Vietnam War, and then once again when the Star Wars pork barrel arrived, was that "If I don't go, someone else will," or "If I don't take the money, someone else will." I call this the replaceability excuse: the argument that if one opts out, one will surely be replaced by someone else who will do the same work, and consequently that it makes no difference to the general outcome whether or not one participates. According to this argument, we can make moral criticisms of the policy choice, but not moral criticisms of those who participate in the operation of the policy, provided that their conduct is consistent with the policy itself and consistent with certain rudimentary moral norms, like the laws of war.

It would be rhetorically unkind of me to mention that the replaceability excuse was offered in the 1950s by staff guards of death camps who were brought to trial, and that the German courts were unmoved by the argument that it was excusable to kill Jews in death camps because someone else was sure to kill them anyway.[9] I will instead offer the more abstract argument that if the replaceability excuse were valid for avoiding the assignment of blame it must also be valid for avoiding the assignment of rewards. But I would be surprised to find a scientist working on a project who would accept the argument that he does not deserve to be paid for his work because, if he had not done it, someone else would have stepped in and done the same work. If the one who does the work deserves the pay, the one who produces the evil deserves the blame, regardless of what others might have done if he had acted differently.

The Small Potatoes Excuse

The replaceability excuse claims that one's contribution makes no difference. The small potatoes excuse claims that in these situations one makes a morally negligible contribution to the bad result. (I used to call this the Eichmann excuse, but I have stopped using this label because I do not want to accept Eichmann's self-characterization as a small potato.) Certainly when one is participating in a large project, it is often difficult to see the impact of one's contribution to the final result. This is especially true when the final result is not actual but probabilistic: for example, an increase in the probability of nuclear war in a given year from 0.005% to 0.006%. But of course being unable to see one's contribution is no argument that the contribution is not there, even if it may be a probabilistic one.

I wish to suggest two different rebuttals to the small potatoes excuse.

The first rebuttal is that in situations in which many people contribute to an outcome, it does not necessarily follow that each bears responsibility for but a small share of the outcome. On the contrary, it often happens that *each* bears moral responsibility for the *whole* outcome. This happens when the work of each is a necessary condition for the success of the project. This condition is more frequently met than people realize. For example, each contributor to a project may mistakenly think that his work was unnecessary because it could have been done by someone else. This is the replaceability fallacy again. If a bit of work is a necessary condition for the success of the whole project—as, for example, designing the exhaust nozzle for the space shuttle is a necessary condition of the shuttle's operation—the person who does that bit of work is morally responsible for the total good or bad produced by the whole project, even if someone else could have designed the nozzle. In general, if action A is necessary for result R, and if it is Jones who is to do A, Jones's action is necessary, even if it could be done by someone other than Jones. Thus I hold each scientist who did necessary work on the

Manhattan Project responsible for the entire expectable outcome at Hiroshima and Nagasaki, even if the scientific work in question could have been done by someone else.

The argument for assigning total rather than partial responsibility for work necessary for an outcome is illuminated by an example from Derek Parfit.[10] Suppose that a building has collapsed and one must choose between (a) saving one person by oneself or (b) participating in a rescue operation with nine others in which five persons will be saved, a rescue operation in which each rescuer plays a necessary role, and for which only you and nine others are available. If one argued that in (a) one person saves one life while in (b) each person saves half a life, then the rational choice of action is (a). But if one does (a) one life is saved and five lost, whereas if one does (b) one life is lost and five are saved. Obviously the correct moral choice of action is (b), and this shows that each of the ten rescuers is morally responsible for saving five people.

The second rebuttal takes in situations in which the project has a built-in redundancy and in which one's work is not a necessary condition of the project's result. In these cases, I believe, each person is responsible for a fractional share of the total harm or good done, and the fractional share may be a very small fraction. But a small fraction of a big result may itself be a substantial good or harm, and it would be incorrect to argue that every small fraction of an outcome is itself morally negligible. Political scientists claim that each voter in a presidential election has a 1 in 100 million chance of determining the outcome of the election.[11] In all other cases, one's vote is redundant. Certainly it would be fallacious to conclude from this that there is no point in voting. Whoever does get elected president makes a substantial difference to the welfare of a great many people. If one believes that military research, on the whole, does harm in the world, or that the military funding of scientific research, on the whole, does harm in the world, the harm is one that affects a great many people, and it is morally quite serious to play even a small part in this business. As a utilitarian will argue, being responsible for a small fraction of an extensive harm can be as serious as being wholly responsible for a smaller local harm. We should be as reluctant to excuse the former as we are the latter.

NOTES AND REFERENCES

1. The connection between moral principles and rational principles is usually associated with ethical theories of the Kantian sort. But the link is also obviously present in egoistic systems and in utilitarian systems, if by "rational" one means the efficient choice of means for maximizing the good. Recently, there have been a flurry of non-Kantian attempts to identify moral principles with principles of rational choice.
2. The overriding character of moral principles is comprehensively argued in BAIER, K. 1958. The Moral Point of View. Cornell University Press. Ithaca, NY.
3. For a recent discussion of universalizability, see HARE, R. M. 1981. Moral Thinking: Its Methods, Levels, and Point. Oxford University Press.

4. For the differences between political ideals and moral requirements see BEITZ, C. 1979. Political Theory and International Relations. Princeton University Press.
5. The main cause of this seems to be the concentration of research on nonmarketable high-tech military products. See Council on Economic Priorities. 1987. Star Wars: The Economic Fallout. Ballinger. Cambridge, MA.
6. At a public meeting sponsored by the New York Academy of Sciences on February 22, 1986, I asked Dr. Gerald Jonas, chief science officer of SDI, what fraction of SDI research could be diverted to an antisatellite mission. Sensible of the congressional moratorium on antisatellite research, Jonas replied "very little." By late 1988, however, satellite destruction was emerging as one of the main missions of strategic defense. See BROAD, W. J. 1988. U.S. promoting offensive role for "Star Wars." N.Y. Times. November 27.
7. For background on these and other incidents, see BLECHMAN, B. 1982. Force Without War. Brookings Institution. Washington, DC. See also PRADOS, J. 1986. The President's Secret Wars. William Morrow. New York, NY.
8. BROAD, W. 1985. Star Warriors. Simon & Schuster. New York, NY.
9. HART, H. L. A. & A. M. HONORE. 1960. Causation in the Law. Oxford University Press.
10. PARFIT, D. 1984. Reasons and Persons: 67–73. Oxford University Press.
11. MEEHL, P. 1977. The paradox of the throwaway vote. Am. Pol. Sci. Rev. **71**.

Strong Role Differentiation
Military Ethics and the Ethics of Military Research

CAROL ANN SMITH

Department of Philosophy
University of Missouri at Rolla
Rolla, Missouri 65401

We live in a time of great changes—changes which require that we commit ourselves to a great deal of rethinking. I refer in particular to changes taking place in the military, and would suggest that a rethinking of the ethics customarily applied to military issues is in order.[a] In this paper, which is part of a much larger project, I will concentrate on sketching the moral situation faced by individuals in the military. I will then turn to the ethics of military research.

Civilian control of the military has certain consequences for military ethics. For example, military personnel acting in defiance of civilian control—whether omitting or refusing to do something they were ordered to do, or doing something they were not ordered to do—would be understood to be in rebellion. Such defiance would threaten the foundations of a democratic nation by raising the spectre of an illegitimate locus of power. Thus, democratic theory, which holds that the public should be the ultimate locus of power, imposes an obligation of obedience on the military. Because often what the military does—harming persons and property—is morally wrong in most contexts, we would expect that only within special constraints could such actions be morally right. If military personnel, individually or collectively, did what they were trained to do, and did so without external constraints, they would be a threat to any public and any nation.

Constraining the military means relying on the duty of obedience, which, however, has serious drawbacks: First, those who have the duty of obedience have also accepted the loss of moral autonomy and moral responsibility. The duty of obedience, thus, conflicts with the claim that moral responsibility is inalienable. Second, those having this duty risk developing a diminished moral sense. Third, there is an enormous potential for abuse, as we have seen in such situations as the Mai Lai incident, in which soldiers followed orders

---

[a] Phenomena such as the world wars we have seen in this century are highly unlikely. We appear to face the devastation of a thermonuclear war or the continuation of so-called low-intensity conflict, that is, a level of international violence either much above or much below that prevailing in the last two centuries.

and committed grossly immoral acts. We thus seem faced with a dilemma: either military personnel strictly uphold their obligation to obedience, and we all accept the attendant risks, or we moderate this obligation and risk the loss of civilian control.

A development in professional ethics can mitigate these drawbacks, I believe. The situation described above appears to satisfy the conditions for what Goldman calls strong role differentiation. Considering the position of judges, Goldman argues that they have a "moral duty to accept institutional obligations to law . . . independent of the content of its demands in specific applications."[1] It is a judge's responsibility to apply the law, not make it, and judges, in accepting their institutional role, bind themselves morally to applying the law, even if they believe that the law is immoral and even if they are correct in that belief.

They may thus, being bound by their role, do something morally right that would be considered morally wrong if done by someone not sanctioned to perform the judicial role. A judge performing in his or her role as a professional is morally right, Goldman argues, if the institution being served fulfills a vital role in society, if observing the professional norm is necessary for the institution to function, and if the professional norm is worth the moral price to be paid. Surveying the claims for strong role differentiation in judicial, legal, medical, and business contexts, Goldman argues that only the judicial context can justify such strong role differentiation. The law is a vital institution in society, and chaos and a lack of stable expectations about what is legal or illegal would result if judges used their moral perceptions on an individual basis. These important concerns justify the strongly differentiated role status of judges, who must fulfill their moral obligations to apply the law even if they believe the law is morally wrong, and even if they are correct in this belief.

In a like manner we may argue as follows: provided that a nation has the right of self-defense, that this right requires a military force, that this military force should be under civilian control, that civilian control is best exercised within the military force by commands coming down from the top (with various appropriate decisions being made at each level concerning means and proximate ends), and that the potential loss of moral sensitivity attending the norm of obedience is worth the moral price, then individuals in the military are in a position where their role is strongly differentiated.

As Goldman argues, and I agree, each institution and profession is unique, and separate arguments must be constructed for each. Obviously there is much more to be said here, both in terms of justifying the conditions listed above and in a closer examination of how strong role differentiation is to be applied. With regard to the application of strong role differentiation, two issues must be immediately addressed: there are international laws that apply to the military, and there are conditions, surely, under which the norm of obedience should be overridden.

The military has dealt with the first issue by interpreting the obligation to obey as the obligation to obey *legal* orders. Legal orders are to be understood as those in conformance with the Uniform Code of Military Justice, with international treaties such as the Hague and Geneva Conventions, and with the customary law of war.[2] The second issue needs a careful series of analyses concerning when the norm of obedience can be overridden. Strong role differentiation places a higher value on obedience than normal moral perceptions would, but strong role differentiation need not place an absolute obligation of obedience on the military. However, this leaves some important questions unanswered: What can the norm override and in what kinds of situations? When can the norm itself be overridden and for what kinds of reasons? The military person is going to have to grapple with these issues, just as the rest of us have to grapple with complex, messy, and difficult issues of morality. It is this possibility of overriding the norm of obligation that preserves moral autonomy and moral responsibility and serves to mitigate the first of the drawbacks described above.

Goldman argues that those who assume strongly differentiated roles risk developing moral insensitivity. Military personnel, however, in moving away from treating blind obedience as a virtue and moving toward a standard of obedience that is less than absolute, and in dealing with the questions noted above, are answering a call for *increased* moral sensitivity and understanding. The norm of obedience is one norm among many. It is weighted much more strongly than such a norm could be justifiably weighted in a civilian context, but it is not absolute, overriding everything else. The attempt to understand and see this clearly is, I believe, a significant step forward and a situation that calls for extremely careful and sensitive moral judgment in those who serve the nation.

Understanding obedience as a strong role differentiated norm preserves moral autonomy and moral responsibility while recognizing the potential limitations on individual action. Like the obligatory force of contracts and promises, it limits individual action but preserves the autonomy an individual needs to reject the limitation. Contracts, however, do not legitimate immoral behavior, whereas strong role differentiation can. Unlike contracts and promises, the norm of obedience is *not* void if immoral. In this respect, the norms of strongly differentiated roles are like laws, which are also not void if immoral. Like a citizen's obligation to obey the law, there is a moral presumption in favor of obeying the norm. There can be circumstances, however, in which the norm can be overridden, such as in cases of civil disobedience. It is not true that "if the King's cause be wrong, our obedience to the King wipes the crime of it out of us."[3] Obedience in the military is not absolute, nor is it a reason for total abdication of moral autonomy and moral responsibility. Nor is it true that obedience can be justified, as it often is, because it has been promised (an oath of office has been taken) or because it is necessary for the military to function.[4] The norm of military obedience is stronger than that because it is grounded as an institutional norm, that is, as a strong role differentiated norm.

Can the norm of obedience in strongly differentiated roles be extended to scientific researchers who are members of the military? Is there a conflict between the military norm of obedience and the oft-cited scientific norms of autonomy and independence? Do scientists have strongly differentiated roles? In what follows, I will blur the distinctions between pure and applied science, between basic and applied science, and between research and development. The important distinction for our purposes here is between directed and autonomous research.

Three claims have usually been made for the professional independence of scientific researchers: 1) Science fosters disinterested, even moral, habits of thinking, and enobles and purifies the mind. 2) Scientists, who are at once expert and morally objective can be counted upon to supply the nation with indispensable disinterested advice and guidance. 3) Science eventually pays rich dividends, in the technology of material wealth, better health, and national defense.[5]

These claims are consequentialist claims: pursuing science in an independent manner will produce more good consequences than if science is externally directed or controlled. There is no claim here that science performs a vital function for the public such as the judiciary or the military. Nor is there the claim that undirected science is a necessary condition for this function. It is claimed that the best results will arise from autonomous science. Although we may recognize this claim to some extent, it has not been regarded as overriding, and we have, in our public policies today, found a place for both autonomous science and science directed toward certain ends. I conclude that scientific independence is not a norm of a strongly differentiated role.

Thus, for the scientific researcher who is a member of the military, I conclude that there is no conflict between two competing norms, and that the military norm of obedience holds sway. The military researcher *qua* military person has a duty of obedience. We need to be much clearer about the details of the application of this norm, however, before the moral obligations of the scientific researcher in the military can emerge with much clarity. One of these details concerns the context in which obedience is required. The heat of battle is one kind of context, the scientific laboratory is quite another. If one stopped for careful moral discussion in the former context, irreversible events would rush by. In the latter context, where time pressures are not as great, the role of the client—the public, the nation—can and must receive more careful analysis and thought. Considering these issues here, however, is not possible.

Scientists who do military research but are not in the military—civilians working for military laboratories, academics performing contracted military research, and corporate scientists working on military projects—are not bound by military norms of obedience. Obedience is *not* a moral norm or a moral virtue in civilian contexts, and contracts to perform morally wrong acts are void. Contracts do not generate an absolute duty of compliance. Civilians entering into employment agreements with military laboratories or corporate scientists and academic researchers entering into contracted military research retain full moral autonomy and full moral responsibility. They are as fully responsible morally for their work as they would be in any other situation. Moral responsibility is inalienable. It may be mitigated by ignorance or coercion, for example, but it cannot be avoided.

I must hasten to add that questions of moral responsibility for military research are greatly complicated by secrecy, which enters, for example, into the conduct of classified research and into the practice of granting security clearances. Although I cannot address this issue at length here, I do want to point out three complications introduced by secrecy.

First, secrecy separates research from normal scrutiny. Guidelines need to be developed to govern science that escapes that normal scrutiny. Second, we have conflicting views of the role of knowledge that need to be thought through. On the one hand, we have a long tradition of regarding knowledge or truth as setting us free. This view regards knowledge as a good for its own sake, a good for human beings. On the other hand, we equate knowledge with power, and hold that the accumulation of knowledge creates power for those who have the knowledge: the power to control access and to manipulate those without access; the power to direct events; the power to reap benefits for oneself in comparison with others. Secrecy interferes with the former view of knowledge by restricting its dissemination and thereby increases the possibilities of the latter. Third, because secrecy may be used to explain ignorance, and because ignorance may be used as an excuse to mitigate moral obligations, secrecy could be used as an excuse to evade responsibility. Secrecy measures could even be used to manipulate the distribution of knowledge such that responsibility could be effectively denied. But these are topics for another paper.

REFERENCES

1. GOLDMAN, A. H. 1980. The Moral Foundation of Professional Ethics. Rowman and Littlefield. Totowa, NJ.
2. Military Qualification Standards I. 1984. Training Support Package: The Law of War (U.S. Army instructional material). U.S. Government Printing Office. Washington, DC.
3. SHY, J. 1988. The American military profession. *In* The Professions in American History. N. O. Hatch, Ed. University of Notre Dame Press. Notre Dame, IN.
4. STROMBERG, P. L., M. W. WAKIN & D. CALLAHAN. 1982. The Teaching of Ethics in the Military. Institute of Society, Ethics, and the Life Sciences. Hastings-on-Hudson, NY.
5. KEVLES, D. J. 1988. American science. *In* The Professions in American History. N. O. Hatch, Ed. University of Notre Dame Press. Notre Dame, IN.

Ethics and Nuclear Weapons Research[a]

PAUL S. BROWN

Defense Systems Program
Lawrence Livermore National Laboratory
Livermore, California 94550

Nuclear weapons are the ultimate weapons. Hence, for many who would debate military research and development, research on nuclear weapons represents the ultimate point of contention. Those of us involved in nuclear weapons research are frequently asked why we do what we do, rather than participating in the supposedly more peaceful endeavors open to scientists and engineers. There are a variety of answers to this question.

MOTIVATIONS FOR NUCLEAR WEAPONS RESEARCH

Most weapons scientists with whom I associate pursue their careers as much out of a sense of doing something important for national security as of having the opportunity to solve extremely complex physical problems. Our work is challenging and large in scale. Working with first-class facilities and first-class colleagues, we have been given tremendous responsibilities in assignments. And, in fulfilling our assigned responsibilities, we eventually realize that we have acquired an added and tremendous responsibility—a responsibility for weapons that, if used, could affect the survival of civilization as we know it.

Nuclear weapons development entails multidisciplinary applied research. The theoretical and experimental problems are among the most difficult in the field of science. Physicists work closely with materials scientists, computer scientists, and engineers to translate concepts, mere ideas on paper, into complex objects capable of being manufactured, incorporated into a military system, and—one would hope—never used.

WEAPONS RESEARCH AND THE NATIONAL POLICY OF DETERRENCE

Why work on something that is never used? The answer is that nuclear weapons *are* used in a very direct way to implement the national policy of

---

[a] This work was performed under the auspices of the U.S. Department of Energy by the Lawrence Livermore National Laboratory under contract W-7405-ENG-48.

deterrence. There is a wide spectrum of views as to what deterrence really is—the views range from minimum deterrence to what some of us like to call dynamic deterrence.

In minimum deterrence, as described by Robert McNamara, the nation should maintain the minimum nuclear force capable of inflicting overwhelming harm on an aggressor who attacks us with nuclear weapons. The threat of mutually assured destruction is sufficient to deter:

> The ultimate goal should be a state of mutual deterrence at the lowest force levels consistent with stability. . . . I know of no studies that point to what that number might be, but it surely would not exceed a few hundred, say five hundred at most.[1]

Dynamic deterrence is like a giant chess game between the superpowers, with moves and countermoves in the strategic relationship being made continually. The idea is to deter war, nuclear or conventional, and to minimize the level of any possible conflict. To accomplish this, as Andrei Sakharov has suggested, requires a continuous modernization of our strategic arsenal, in order to respond to events that impinge on its effectiveness, and hence, its deterrent value:

> While nuclear weapons exist, it is also necessary to have strategic parity in relation to those variants of limited or regional nuclear warfare which a potential enemy could impose, i.e., it is really necessary to examine in detail the various scenarios for both conventional and nuclear war and to analyze the various contingencies.[2]

Anticipating different contingencies so as to maintain the strategic balance is where weapons laboratories fit in. We view the goal of our research as avoiding war, rather than preparing for war. The ultimate goal is to avoid mutual destruction. The ethical consequences of our work should be viewed in light of this goal.

Each year that passes without a war between the superpowers adds to the evidence that deterrence is working. This is most reassuring to those of us who work to maintain the nation's deterrent.

ETHICAL IMPLICATIONS OF NUCLEAR WEAPONS RESEARCH

Do we, as weapons scientists, think about the implications of what we do? Very much so. It has been difficult to avoid such thoughts in the face of the continuing pressures against weapons research, particularly since the early 1980s, with the advent of the nuclear freeze movement. My own thoughts become focused each time I have to pass through the barriers of demonstrators who periodically parade in front of the laboratory gates. I believe that instead of dissuading me and my fellow researchers from continuing to do what we do, the demonstrators have caused us to consider the ethics more thoroughly. The result has been a reinforcing of personal convictions as to what our job is all about. In this sense, I believe the demonstrations have been a good thing. They have heightened consciousness and have served to remind people about some very fundamental issues.

The demonstrators actually represent a small minority, whose numbers have been declining over the years. Over 1200 out of 3000 were arrested at a demonstration at Livermore in 1982, and today we rarely see even 100 show up at the front gate. The true majority in the country is the one that has voted in the national elections. People have generally expressed their wishes for a strong national defense at the ballot box. It is more complicated than wanting peace through strength. As Joseph Nye has put it, the American public wants peace *and* strength, hence, the largely centrist positions that have evolved in political campaigns and in the U.S. Congress. If the weapons laboratories represent anybody or anything in the country, I believe it is the largely centrist position that exists within the government. In truth, this is more of a populist than an ethical justification.

I believe that we as weapons scientists think a lot more about the policy issues than most people in the general public. After all, these issues are part of everything we do. We constantly reason, question, and debate. We react to questions about the morality of our work. Those who demonstrate against us can go home at the end of the day and take up where they left off at the next demonstration. We go behind the gates and continue to think about the implications of our work.

What concerns me most about the demonstrators is a common assumption, noted by Joseph Nye, about our motives:

> Some who oppose nuclear deterrence discount the views of those who defend it as corrupted by "the disease of nuclearism." Instead of meeting their opponents' arguments, they make up a theory about their opponents' motives. They try to shrink their opponents' stature rather than refute their arguments. When they do that, they are involved in political caricature, not in moral reasoning. Without a degree of humility and charity, we are condemned to shouting such caricatures at each other, and the illumination of moral reasoning is snuffed out.[3]

How are ethics relevant to what we do? First, we must do our research in an ethical manner. Ethical behavior is expected of all professionals, scientists and nonscientists alike. There is, however, a difference for scientists who do weapons work because of the existence of bureaucratic and political pressures that are lacking in other fields. Second, ethics help us think about the consequences of the weapons we develop. Is it ethical, we wonder, to even be doing the work in the first place, given the terrible consequences we would suffer if these weapons were ever used? Conversely, what are the consequences of not doing the research? That is, why is the work important? I will discuss these two general subjects in turn.

DOING WEAPONS RESEARCH ETHICALLY

The Influence of the Weapons Laboratories

There has been criticism over the years that the weapons laboratories exercise undue influence in pushing and selling their programs, and serious questions of ethics have been raised. In point of fact, the laboratories have been chartered by the Congress, very specifically each year in the Defense Authorization Act, to explore and provide new technologies necessary to maintain U.S. nuclear deterrent forces and to conduct research on the feasibility of innovative applications of nuclear technology that may eventually be important. An extremely important part of this task is to communicate to the government what can be achieved, and this process of communication is what many refer to as influence.

The influence of the laboratories has been put into perspective by Bryan Hehir, a Catholic priest who was a key architect of the American Bishops' Pastoral Letter on War and Peace (in this letter, the bishops gave a conditional, interim acceptance to nuclear deterrence).[4] In February 1985, at a colloquium sponsored by the Bishop of the Oakland Diocese, Father Hehir replied to a question about the weapons laboratories' role in pushing technology and their impact on dominating policy:

> It seems to me the weapons laboratories are going to do what they are designed to do. They're going to be totally involved in this process. They're going to put forward a whole series of propositions. They're going to try to push forward the frontiers of scientific research, and they're almost inevitably going to push for technological transition. I'm not against people doing what they are designed to do. . . . I'm in favor of . . . political figures doing what they're designed to do. . . . They ought to listen intently to what scientists and technological institutions propose, and then they ought to decide on other grounds than the purely scientific or technological grounds what ought to be done. . . . I'm sure the weapon laboratories do drive the dynamics, but I'm not positive . . . that they ought not to do it. I'm arguing that other people ought to do other things.

In a recent lecture at Stanford University on the future of the nuclear debate, Hehir was asked if people working on weapons should be persuaded to work on arms control and peace-related efforts. Hehir responded that the most important goal is to avoid nuclear war, and this required proper strategic balance. He said it was just as important to have excellent people working on sensible strategic improvements aimed at improving stability as it was to have excellent people work on arms control and peace initiatives. He said we must do both.

Weapons scientists have a responsibility that goes beyond just that of exploring new technologies. There have been advances in weaponry that have overcome tremendous technological challenges, but have also introduced instabilities into the strategic relationship. The development of missiles carrying multiple, independently targetable warheads is one example. It behooves us as scientists to do our best in working on a new technology, and we must also speak out on what is wrong with a technology, particularly when its implementation is apt to introduce serious instabilities. We indeed have been speaking out much more in recent years, and examples are given below.

Technical Objectivity as an Expression of Ethical Behavior

The nuclear weapons laboratories have a broad range of technical missions. In general, we exist to do research and development on nuclear weapons. Our specific missions are to maintain the reliability of the existing stockpile of nuclear weapons; to modernize weapons for improved safety, security, survivability, and military effectiveness; and to maintain the high level of expertise necessary to accomplish these other missions. We also work to determine what is possible for our potential adversaries in nuclear weapons technology. We work to support national objectives in arms control. In fact, arms control is an integral part of everything we do: the same expertise that is used to develop weapons is directly applicable to their control.

Although the government assigns the programs on which we work, and imposes certain conditions that must be satisfied, as individual scientists we strive to maintain technical objectivity in carrying out our work. There is room for improvement, as in any field, but because we are managed by the University of California, we may enjoy our status as University employees. We are able to work in an environment that encourages technical excellence and freedom from political and bureaucratic pressures. University management provides an atmosphere where debate is possible and where intellectual reasoning is dominant. In a different atmosphere, such as government or industry, there would be less enlightened and independent-minded research, which could lead to cruder and more dangerous weapons or the pursuit of poorly conceived concepts.

The following example shows how the existing environment has served to bring reason into the strategic debate. In 1982, President Reagan set up a commission led by Charles M. Townes, a Nobel Prize winner in physics from the University of California at Berkeley, to evaluate basing options for the MX missile. The commission sought advice from a variety of sources, including weapons systems analysts from Livermore. Livermore scientists had been studying various basing options for the MX and had concluded that all suggested basing schemes were flawed. That Livermore should have offered such criticisms was somewhat ironic because Livermore had also been assigned the task of developing the nuclear warhead for the MX. While the right hand was developing the weapon, the left hand was showing what was wrong with the overall system.

Dr. Townes wrote University President David S. Saxon and said the following about the Livermore contributions to his commission's efforts:

> It was clear that most of the industrial organizations were quite cautious about giving information or making conclusions which would be contrary to Pentagon policy. I was personally impressed that the many persons who helped us from Livermore seemed completely objective in examining the technical facts, in investigating what needed to be looked into, in looking for weaknesses as well as strengths in current proposals, and in being willing to state plainly, though diplomatically, where they did not agree. In most cases, I found individuals from academia also objective, though generally by no means so deeply knowledgeable. . . . I make the above point because I think, contrary to some opinions, Laboratory personnel are often important in giving helpful perspective and ameliorating U.S. nuclear policy, and that this is partly because they are protected by the management structure from obvious pressures to which commercial companies or governmental laboratories are subjected.

The Townes Commission study took place seven years ago, but the basing debate goes on. In the winter 1988 issue of *International Security* magazine, there is an article[5] on the rail basing of the MX missile by John Harvey, a Livermore physicist, and Barry Fridling, who was recently an arms control research intern at Livermore and jointly a MacArthur fellow at Harvard's John F. Kennedy School of Government. The article addresses a fundamental problem with the concept of rail basing: the survivability of missiles in rail-based systems depends on strategic warning, something our nation has historically failed to recognize. Although the article runs counter to current Administration policy, it does so with technical objectivity.

Another example where Livermore personnel have come to scientific conclusions independent of external policy considerations has to do with whether the Soviets are in compliance with the Threshold Test Ban Treaty (TTBT). The TTBT, negotiated and signed by the United States and the Soviet Union in 1974, limits the yields of underground explosions to 150 kilotons. Although the treaty has yet to be ratified, both parties have agreed since 1976 to observe its conditions. Monitoring compliance with the treaty is done by teleseismic means. Each country maintains seismometers on its own soil to measure the long-range seismic signals produced by nuclear explosions on the territory of the other side. There are large uncertainties in seismic monitoring of Soviet nuclear test yields (and conversely of U.S. test yields), and the Reagan Administration has claimed that the Soviets are in likely violation of the treaty. Yet, studies by verification experts at Livermore have concluded that the Soviets have been observing a yield limit, and that this limit is consistent with TTBT compliance, although a few tests might have exceeded the limit. We have stated our results publicly, have reported them in testimony by Livermore's director[6] and other scientific staff members[7] to the U.S. Congress, and have had a significant impact on the U.S. policy debate.

Our stance on the TTBT has had its repercussions. In 1983, Roy Woodruff, then Associate Director in charge of the nuclear weapons design program, and Bill Scanlin, Woodruff's deputy, were interviewed by the Washington Post. They expressed the view that they favored the 150-kiloton TTBT limit, that the monitoring evidence because of its uncertainties failed to prove or disprove Soviet violations of the TTBT limit, and that even if the Soviets had violated the limit, the small variances above 150 kilotons would not have given them any military advantage in weapons building. Woodruff reports that at 5:30 A.M., California time, he was awakened by a phone call from the Secretary of Energy asking how he dared oppose Administration policy on this matter. Roy reminded the Secretary that he worked for the University of California and *not* the Administration, and that he had every right to speak his own mind on the issues. Roy reports that the Secretary apologized and immediately invited him back to Washington to discuss the issues, and that these discussions were productive.

There are other examples. Livermore has analyzed the systems requirements for deployments of the Strategic Defense Initiative (SDI). These analyses have often provided a less optimistic basis for projecting what might be accomplished by potential SDI deployments than has been projected by other system analysts or by certain ardent proponents of SDI (both inside and outside Livermore). A good example of this is the recent study of early SDI deployments using space-based rockets. Our analysis has shown that while such early deployments may provide an effective defense against the current Soviet missile threat, ten times as many rockets would be required to defend against a modernized Soviet force. Our analysis also shows that the Soviets could implement cost-effective countermeasures to the proposed early deployments over the same time frame as these deployments. In April 1987, George Miller, the head of the Livermore weapons program, reported on these analyses in testimony to the U.S. Senate.[8]

With respect to the feasibility of SDI in general, there is a wide spectrum of views at Livermore as to what might be achieved. The national consensus seems to be that there should be a healthy research program. My colleagues, however, generally view the idea of achieving extremely effective defenses with healthy skepticism. Still, virtually all Livermore staff members, including the senior program managers, believe that research should be done within the limits of the Antiballistic Missile Treaty (something easily achieved for our SDI research programs). Nearly everyone I know supports the treaty; nearly everyone shares the traditional interpretation of the treaty. All these views have been held in the face of unquestioning support for SDI by the Administration in Washington.

There are those in the nation and at the laboratories who firmly believe in the promise of defensive systems. They view SDI as leading to weapons systems that would be far preferable to those we now rely upon for offensive retaliation against an enemy attack. They argue that it is more moral to build defensive weapons than offensive (or counteroffensive) weapons. It is difficult to refute such an argument. A broader view on this issue was recently expressed by the National Conference of Catholic Bishops in a follow-up to their much-debated pastoral letter.[9] The bishops stated that the moral character of SDI must be judged more on its consequences than on its intent. Indeed, the concerns expressed by our scientists about SDI do address the consequences of that program.

The above examples of independent thought and action typify the technical objectivity that exists among the scientists at the weapons laboratories. This technical objectivity has often run counter to bureaucratic pressures in the government and has been entirely consistent with proper professional ethics. For the most part, we have tried to do what is technically correct, rather than what is politically expedient.

Room for Dissenting Views

Technical objectivity and professional ethics demand that there be room for dissenting views in any research establishment. There are scientists at the laboratories who do question that technical objectivity exists across the board. At Livermore, for example, there are physicists who have been openly critical of how the laboratories do their business. They believe that the laboratories have been overly aggressive in pushing technical programs at the expense of technical objectivity. Specifically, they are highly critical of the approach to designing weapons. Among other things, they have stated that the weapons we have designed for the stockpile should have been more conservatively designed with a comprehensive test ban (CTB) in mind and that the laboratories have used their influence to fight a CTB, intentionally developing weapons that require further nuclear testing to keep them reliable. They have implied that the laboratories place obstacles in the way of achieving a CTB and other arms control measures just to stay in business, and are only interested in developing exotic new weapons like those for SDI.

In 1987, the director of Livermore agreed to a request made by six congressmen to make Ray Kidder, one of the scientists, available to do a study on stockpile reliability.[10] We provided Kidder with the information he requested to do his study. A corresponding study was simultaneously done under the direction of the head of the Livermore weapons program.[11] These events—which, I believe, mark the first time that an internal critic was asked to write a dissenting study—show how much freedom there is at Livermore to express alternative points of view.

The dissenting scientists have exerted a lot of leverage on the debate. Although I disagree with what they have been saying, I find it reassuring that they are free to express their ideas, and to criticize, because of the University management of the laboratories.

Risks of Technical Objectivity

There have been news stories about disputes between Roy Woodruff and Livermore management concerning overly optimistic assessments for the X-ray laser made by Edward Teller and Lowell Wood to highly placed policy makers. Roy resigned his position three years ago because of these disputes. A key issue then was who had the right to speak for Livermore's X-ray laser program—a program Roy was in charge of. Roy wanted to write letters to the same policy makers to present his and the program's views about the technical possibilities, but he was told by management[12] that they preferred he not write those letters, and instead brief the policy makers in person. Roy disagreed with this management decision and resigned.

Roy set his principles high. In so doing, he put his extremely promising career in jeopardy. What we are dealing with here is not just a matter of who was right and who was wrong about the technical assessments. There have always been disagreements among scientists in the past, and there always will be in the future. What concerned Roy was how the assessments got carried forward and influenced the policy debate. I believe he was right to be concerned.

Now some have used this example to discredit the X-ray laser program. It is important to note, however, that Woodruff has been a strong supporter of the X-ray laser research at Livermore and has stated so publicly. The critics of a program will always use whatever publicity is available to their advantage. This is a risk of being technically objective. I believe, however, it is a risk we must live with.

THE ETHICS OF DOING NUCLEAR WEAPONS RESEARCH AT ALL

Having discussed doing weapons research in an ethical manner, I now return to the question of why it should be done at all. Nuclear weapons are not unique when it comes to ethical considerations of weaponry and war. The debate about what is just warfare has gone on for centuries. Nuclear weapons do add a new dimension to the problem, considering the threat they pose for the survival of civilization. War can no longer be restricted to combatants, and innocent civilian populations are now vulnerable to annihilation.

How do we in the weapons laboratories justify the role we play in developing weapons that could have such dire consequences for civilization? I'll give you my view, which I believe is shared by many of my colleagues.

There are some scientists for whom the technological fix is almost a religion. Fortunately, I have found few such scientists at Livermore. Perhaps we who appreciate the limits of technology are less willing to put complete faith in what it can do. Take SDI as an example. From the surveys and polls I have seen, it appears the general public is much more optimistic about SDI than those in the scientific community, including those of us in the weapons laboratories, as borne out by the examples given above.

We also realize the limitations of nuclear weapons. They exist to deter, and their use implies failure of other political and military approaches. I personally believe that the solutions to today's problems will have to be political and social. The role of technology is to help us to survive until we achieve the necessary political and social solutions, to "buy time" as one of my close colleagues puts it.

Joseph Nye, in discussing the moral pros and cons of nuclear deterrence, has endorsed the morality of a nuclear deterrence that properly accounts for the risks of various policies and their alternatives, and has addressed the "means" of deterrence:

> . . . if there is absolutely no possibility of the use of nuclear weapons, or if that is believed to be the case, they will have no deterrent effects. Thus deterrence depends on some prospect of use, and use involves some risk that just war limits will not be observed.[13]

The weapons laboratories work to develop the means of deterrence, to improve them constantly, and to help minimize the risks associated with deterrence policy. Although many accuse the weapons laboratories of fostering a war-fighting mentality, we strongly believe that developing weapons that are survivable, safe, secure, and effective enhances the means of deterrence and reduces the risk of war. Perhaps it is developing weapons that are thought of as being effective against real targets that causes the most concern about warfighting. It is good to keep in mind what Nye has said in this regard:

> Thus credible targeting seems necessary for deterrence. Yet the effort to identify credible military targets has raised a heated debate over the legitimacy of planning for "war-fighting" as opposed to nuclear deterrence. In one sense this debate is spurious, for some planning for the delivery of nuclear weapons against military targets is both planning for "war-fighting" and a necessary means of deterrence.[14]

The work of weapons scientists supports the dynamic deterrence concept described above, and is vital to that form of deterrence. The consequences of not doing the kind of work we do would be greater instability in the world and a weakening of a policy that has worked for over forty years.

The weapons laboratories have taken a share in arms control efforts over the past few years, and our contribution to these efforts is growing. The same knowledge that is used to develop nuclear weapons can be applied to their control. Arms control is an integral part of everything we do. Arms control has also done something for the weapons laboratories in providing an endeavor that is more acceptable to those who criticize weapons development activities.

Arms control is just one political process that can help us reach a better world. Arms control can serve a variety of purposes, such as increasing stability, saving money on expensive weapons systems, reducing the damage that might occur in war, lessening the threat each side poses to the other side, and reducing the risk of accidental war. It has done all of these things with varying degrees of success, as well as something even more important. As a political process, arms control gets the two sides talking to each other, working together to reduce tensions and build a better relationship. We are at a propitious time in the history of arms control, with much progress over the past year and more to be made in the near future.

Getting Ethics into the Weapons Debate

Although we can justify to ourselves the ethics of our work, there are still those who question those ethics. I become concerned when I hear people question our motives, and I feel compelled to communicate with the questioners and convince them of the importance of our work, as well as of its ethics. I am concerned because I believe that my goals and the goals of the questioners are really the same. We simply have different approaches to solving the problem. Michael M. May, an Associate Director at Livermore, said it very well in an op ed piece entitled "Demonstrations: A Moral Equivalent of War" that appeared in the *Christian Science Monitor* on December 29, 1986:

> Wittingly or not, the demonstrators are doing much the same thing as the U.S. government does in maintaining a nuclear deterrent: putting on a show of war to avoid a much worse war. Of course, the stakes are different. But the range of feelings and the range of attitudes are not so different. The relation of either activity to peacemaking is the same.

Because we are concerned, we communicate. Maintaining a dialogue on arms control is one of the most important things we can do.

The Bishop's Colloquia on the Ethics of Deterrence

Joseph Nye has commented on the difficulty of maintaining a dialogue between the different specialists who are interested in nuclear weapons issues:

> All too often moralists and strategists tend to talk past each other as though they lived in separate cultures of warriors and victims rather than fellow citizens of a democracy. The moralists formulate fine principles that seem to the strategists about as relevant to foreign policy as a belief in the tooth fairy is to the practice of dentistry. The strategists, on the other hand, tend to live in an esoteric world of abstract calculations and a belief in a mystical religion called deterrence, which is invoked to justify whatever is convenient. Strategists would do well to realize that there are no experts, only specialists, on the subject of nuclear war, and to listen more carefully to the moralists' criticisms. At the same time, philosophers and moralists would do well to pay more heed to the strategists' arguments and to realize that they will need to work with more realistic assumptions if they wish to be effective in a dialogue between ethics and strategy.[15]

The participants in at least one attempt at a dialogue seem to be taking note of Nye's advice. Four years ago, members of the San Francisco Bay area religious community and scientists at Livermore and the University of California started taking part in a series of colloquia inspired by the Bishops' Pastoral Letter.[4] Several scientists at Livermore who are Catholics spoke with John Cummins, the Bishop of the Oakland Diocese, and pointed out that the presence of two nuclear weapons laboratories (Livermore and Sandia), a large university, and a major theological seminary in the Diocese made the area an ideal place to have a dialogue on the issues raised by the Pastoral Letter. The Bishop was convinced and began funding the colloquia. A few years ago, the University of California Institute on Global Conflict and Cooperation took over the funding of these events.

These colloquia have had a host of distinguished speakers from diverse backgrounds, including defense policy, defense research, political science, pacifism, ethical philosophy, and religion. There has been considerable interchange between the speakers and the participants, and between the participants themselves. Although we have frequently talked past one another, we have begun to establish a rapport with each other: listening to each other's concerns and thoughts has made us more willing to appreciate each other's perspectives.

Initially, there was concern that sparks would fly by convening such a seemingly disparate group of individuals. This has yet to happen. Our success may in part be due to our having excluded the press in order to remove barriers to open expression and to avoid grandstanding by those who might be so inclined. The seventh colloquium was held at Livermore and was attended by almost one hundred people. Its spirit was very well expressed by Michael May, who in his welcoming speech said, "We have all come here to worry about nuclear weapons together."

We have discussed a broad variety of subjects: the Bishops' Pastoral Letter and its intent; the future of nuclear deterrence; alternatives to deterrence; whether the Pastoral Letter did too little, was about right, or went too far; deterrence in the next ten years; nuclear testing and how it relates to the mission of the laboratories; and, most recently, the ethics of deterrence. We are beginning to find that those taking part in the dialogue are getting closer; people are getting many more direct answers to the questions they are asking— even though some of the questions are indeed quite tough and strike right at the heart of the matter. I would like to see our policy makers exposed to such a dialogue. It might make very little difference, or it might at least make them think about weapons issues in a different context.

GETTING MORE ETHICS INTO THE PROCESS

The question remains as to how we might instill more ethics into policy decisions. At a recent seminar at Stanford, Bryan Hehir was asked if the debate about ethics could make a difference. He replied that in order for the debate to have an effect, people like himself would have to continue to push, would have to make themselves "more convincing," and that he intended to keep pushing. He said that people generally perceive the role of ethics in one of four ways: ethics is superfluous, it is decisive, it is corruptive, or it is complementary. I believe that ethics should play a complementary role to sensible strategic planning, which itself should be based on an ethical policy.

I believe that ethics will play an ever-greater role in the nuclear debate in the future. As weapons scientists, we will continue to have the task of designing the hardware that forms the means of deterrence. Although we will do the best technical job possible in developing that means, we must continue to be prepared to say what is wrong or destabilizing about that means.

REFERENCES

1. McNamara, R. S. 1986. Blundering into Disaster. Pantheon. New York, NY.
2. Sakharov, A. 1980. Foreign Affairs 59(2).
3. Nye, J. 1986. Nuclear Ethics: 11-12. Free Press. New York, NY.
4. National Conference of Catholic Bishops. 1983. The Challenge of Peace. U.S. Catholic Conference. Washington, DC.
5. Harvey, J. & B. Fridling. 1988/1989. On the wrong track? An assessment of MX rail garrison basing. Int. Security 13(3): 113-141.
6. Batzel, R. 1987. Testimony before the Senate Armed Services Committee. February 26.
7. Nordyke, M. 1987. Testimony before the Senate Armed Services Committee. February 26.
8. Miller, G. 1987. Testimony before the Senate Appropriations Committee. March 26.
9. National Conference of Catholic Bishops. 1988. Building Peace: A Pastoral Reflection on the Response to "The Challenge of Peace." U.S. Catholic Conference. Washington, DC.
10. Kidder, R. 1987. Maintaining the U.S. Stockpile of Nuclear Weapons During a Low-Threshold or Comprehensive Test Ban. UCRL-53820, October.
11. Miller, G., P. Brown & C. Alonso. 1987. Report to Congress on Stockpile Reliability, Weapon Remanufacture, and the Role of Nuclear Testing. UCRL-53822, October.
12. General Accounting Office. 1988. SDI Program—Accuracy of Statements Concerning DOE's X-Ray Laser Program. Briefing Report to the Honorable George E. Brown, Jr., House of Representatives. GAO/NSIAD-88-181BR, June.
13. Nye, J. 1986. Nuclear Ethics: 52.
14. Nye, J. 1986. Nuclear Ethics: 105.
15. Nye, J. 1986. Nuclear Ethics: 10-11.

The Ethics of Corporate and Academic Commitments to Military Research

JOSEPH F. COATES

J. F. Coates, Inc.
Washington, DC 20015

This paper is metaethical. My conclusion from a metaethical point of view is that using ethical terms in discussions of corporate and academic participation in military research is pointless at best, and is in fact more likely to be counterproductive. With that as my conclusion, let me build the case for it.

First of all, I think we can agree that almost all technologically and scientifically driven activities of our society are loaded with ethical considerations. Military affairs are more heavily burdened with ethical and moral concerns than most other areas. But that is a thin excuse for framing a discussion in ethical terms. It would be much like saying that because physics is so much an essential part of our transportation, our energy, our groceries, our movement, our day-to-day activities, every discussion should begin or end with physics. Simply because a factor permeates a subject is itself no excuse for talking about it in terms of that factor.

But, we must look further for reasons to avoid discussing these institutional dyads—the military and the corporation, the military and the university—in ethical terms. The ethics of our society, whether formally religious or merely secular, are fundamentally Judeo-Christian, and so are based on moral beliefs that are personalistic. The ethics in our society are framed around what you should do in relation to other people, what you should do in relation to yourself, what you should do in relation to institutions, and (for many) what you should do in relation to the Divinity. Yet the kind of ethical questions raised by the conduct of military research have almost nothing to do with us as individual practitioners in the corporation or in the university. The fundamental questions are of relations of institutions to institutions. Our Judeo-Christian culture is bereft of any significant ethical theory, discussion, or framework relating an institution to an institution. It is much as if one were attempting to talk physics before Aristotle, much less to talk it before Newton. There just is not the basis for the kind of discussion that many of us would like to engage in.

For those who would reject this notion that there is no established ethical framework for institution-to-institution relationships, we must turn to other aspects of this personalistic, Judeo-Christian heritage. Ethical reasoning is a formal activity. To behave ethically as individuals and to see ourselves as ethical beings does not mean that we have mastery or control of ethical paradigms, thoughts, or frameworks. Rather, our ethics on a day-to-day level are the ethics manifested in our behavior, not in our cognitive processes. The subject at hand of course is basically one that calls for cognition, not for individual day-to-day behavior. To put it differently, most of the people who would be concerned with the subject of corporate or academic relations to military research are untrained, undisciplined, and unpracticed in ethical reasoning. In my judgment, the autodidact from time to time comes forth as an effective ethicist, but by and large, ethics is not a subject to be comfortably left in the hands of amateurs. Furthermore, it is clear that a majority of people most immediately concerned with the subject at hand—school administrators, corporate officials, corporate executives, and so on—have had little or no formal training in ethics.

For most of us, ethical training stopped somewhere in the teenage years, where the primary focus was on the concerns of teenagers, social/sexual relations, family responsibility, response to personal authority, and so on. For most of us our ethical and moral training stops well before we are twenty years old. So what we have are inappropriate tools in the hands of people even in the limited personalistic framework. Equipped with the wrong tools, many people attempt to address the question of corporate and academic ethical roles. In the absence of an ethic of the institutional collectivity, the overwhelming number of relationships between the military and the corporation or the university academic are without any basis for discussion. As I see it, ethics has become a taxonomic catch-all, the place where we dump all those knotty problems that we have not had time to think through. Put in the starkest terms, most of our individual responses to ethical questions are barely more than a response to a bubble in the belly. They have no intellectual, cognitive, systematic, or moral basis, but rather reflect some kind of animal disquiet that we label, for want of a better term, as ethical.

We are confronted in America by further confounding factors in our ethical discourse. The United States has two characteristics that are pernicious when it comes to the public discussion of ethical questions. An alarmingly large percentage of Americans are moralistic in the most narrow, petty, parochial way. I recall, for example, when I was working for the Institute for Defense Analyses, at a meeting very much like the one upon which this volume is based, I was in a group standing around making mutual introductions. A not-so-young man, perhaps at that time thirty, refused when I reached out to shake his hand. I thought that was the most pitiable kind of thing. The refusal to shake hands could not have been based on even a full sentence of information about me. What he was doing was reflecting the same kind of mentality that says "Jews are hateful," "Niggers are dumb," "Japs are our little yellow enemies." That type of categorization and categorical thinking was prepared to wipe me out of the galaxy on the basis of knowing absolutely nothing about me but my employer's name. That kind of moralism is infectious and pernicious in American society. We must fight it, not feed it.

The second characteristic of our society, political diversity, in contrast to moralism, is a strength, but not one facilitating ethically grounded debate. There is little or no hope for a broad basis of agreement in American society on core moral or ethical beliefs. We are heterogeneous and becoming more heterogeneous. The likelihood of any shared moral beliefs is diminishing. The hope of bringing us together encourages two pathologies of the contemporary ethicist.

One of these pathologies is the search for values clarification. The last thing in the world I want is for your values to be clarified publicly. Why? Because the chances are that if your values were clarified, they would be falsified or hypocritical, because in clarifying your values you are not likely to reveal you hate this or the other group, or that you have other morally dark values. What you are going to do is give a goody-goody middle-class gloss to your beliefs. Values clarification is likely to be both fraudulent and unachieveable. Even if we could clarify our values publicly, doing so would provide us with no direction, only the basis for further discussion.

The other pernicious element that comes from the attempt to heal the democratic diversity in our society is the advocacy of situational ethics. The situational ethics route is itself destructive. Why? Because with regard to the kinds of issues of concern in dealing with the military, those who would speak from a situational point of view are almost always parochial. The situational argument cannot embrace us all; quite the reverse, it is extremely narrow and personalistic.

Frustrating any attempt at establishing an acceptably broad ethical base in our society are rapidly shifting social/secular values. For example, in the lifetimes of more than a few people who heard me present this paper, Louis Fieser was celebrated for his accomplishments in developing napalm. Fieser, a Harvard professor, developed napalm during World War II specifically to deal with the Japanese, who were accepting lethal casualties, that is, deaths, among 98–99% of their people in the bunkers. Napalm was developed to burn them out. Now, napalm has not changed. What has changed is the social context in which we look at weapons. Anyone who would address the question of whether Fieser was ethical could hardly cope with the question of changing social values. Presumably, any institutionally oriented ethical framework dealing with the military ought to be applicable for more than a mere forty years.

It is unlikely many people gave much thought to how napalm might be used after Japan was defeated. Those who welcomed the introduction of napalm probably did so out of patriotism. Thus the example of napalm raises another point: not everyone can think everything through. How can we expect the run-of-the-mine academic, the run-of-the-mine corporate executive, or the run-of-the-mine corporate researcher to be able to think through with crystalline clarity every putative ethical issue of his or her work? There are so many complications, from every point of view (for example, there are political, economic, and technological side effects), that in most instances we must resort to some other way to make decisions—patriotism, confidence in leaders, a semiauthoritarian model, or a long-term vision about society, its health and its enemies.

At the level of the individual, another critical point should be noted. Raising ethical issues does not necessarily promote understanding, and could in fact create barriers to understanding. Raising ethical issues almost always implies that at least one of the actors in a discussion must acknowledge the implicit or explicit charge that his or her actual or intended behavior is unethical, immoral, or at least morally questionable. Such an acknowledgment is extremely difficult for most of us, and hence the issue shifts from the topic of concern to the individual's ego and self-image.

Now having cleared the underbrush, we must face the question of what do we do when confronted with these putatively ethical issues. We have to have a point of view on who it is who is dealing with the issue. Let us look at it from the point of view of a corporate executive, a corporate manager, a university administrator, or a department head, not from the point of view of the individual employee. The healthy, effective, and fruitful way to deal with these issues, from a functionary's position, is to regard all partisan claims and points of view as mere individual preferences, and to run a calculus in terms of the most crass self-interest for the institution. For example: If we support project X, what will it do to our student recruitment? If we support project X, will it disrupt the campus? If we support project X, will it cut foundation money? If we support project X, will it lead to a cutback in state allocations?

If one expects functionaries to do no more than serve their institutions' interests, however narrow these interests may seem, one can in fact sharpen discussions and come to rational and significant conclusions. I advocate that one abjure, skirt, and pay as little attention as possible to attempts to resolve ethical issues; instead, one should treat ethical points of view about an institution's behavior as mere data about ideas held by the institution's constituents.

To illustrate the point in a nonmilitary institutional context, suppose you are a bureaucrat involved in some decision that deals with the conflicting ethical values of "pro-life" and "freedom of choice" advocates. As a citizen, the last thing in the world I would want is for some bureaucrat at a university, at a research establishment, at a corporation, or at a government agency, to make a moral judgment or decision between those two extreme positions. What I would hope is that the bureaucrat would weigh the forces and come to some minimum pain/maximum payoff solution in terms of the organization's interests. That would be rational, and it would be a step forward. The kinds of considerations that should enter into what one would choose to consider ethical considerations, I would boil down to effects on reputation, effects on funding, effects on long-term institutional viability, side effects on hiring, side effects on employment, and so on.

One could counter what I have said, arguing that we will never get to dealing with issues ethically if we do not start to deal with them ethically. That is certainly a rational and plausible point, but we must not kid ourselves that because we have a bubble in the belly, we have a rational basis for discourse. Instead of arguing on the level of individual ethics, which do not apply, we should concentrate on developing institution-to-institution ethics. A long-term effort will be needed to develop these ethics, however. What this paper addresses is what to do in the meantime.

Let us consider the pseudoethical elements. Right now the Department of Defense (DOD) is putting forward ethical codes to be adopted by contractors. These codes are a farce. What these codes effectively make are statements of good behavior. Companies will adhere to these statements of good behavior for only three reasons. If you violate them, you may get thrown in the slammer. If you violate them, your company may not get any more DOD business. If you violate them, your reputation may suffer to a degree that influences non-DOD business.

These are down-to-earth considerations that have nothing to do with thinking in ethical terms. One should just adhere to the rules. If one adheres to the rules, everything is okay. The behavior of an organization almost always results from the organization's having been subjected to extreme pressure. Organizations knuckle under to the pressure eventually. Having knuckled under, they develop a code or a set of rules and guidelines that puts a good face on the unavoidable. But no organization that I know of has ever developed a code without external pressure.

Corporate managers find that codes are valuable when somebody raises an issue that is questionable from a business point of view. That is, codes may be used to reduce the amount of useless discussion.

Consider animal research in connection with the DOD. It does not make any difference to me what one's position is on animal rights. If I were a DOD official, my concern would be where the power is, what the consequences are, what the alternatives are. I do not want any official in the DOD saying animals do or do not have rights. I do not want anyone in DOD drawing that kind of moral judgment. Yet many would relish ensnaring a DOD official in that mare's nest.

My metaethical analysis suggests that the only moral failing that any of us might routinely experience is acting in ignorance. This is quite a different matter from consciously acting unethically or immorally.

Ethics and Biological Warfare Research

LEONARD A. COLE

Department of Political Science
Rutgers University
Newark, New Jersey 07102

Biological weapons, unlike any other weapons, have been entirely renounced, and their possession by nations has been prohibited by international treaty. Nevertheless, many ethical questions that apply to biological warfare research have bearing on other forms of military research. This is especially true where specific categories of arms have been or are expected to be restricted by law. In this regard, the relationship of scientists to the U.S. Biological Defense Program can be instructive.

BIOLOGICAL WARFARE IN PERSPECTIVE

Biological weapons have often been coupled with chemical weapons, insofar as both are seen as particularly repugnant and less controllable than other instruments of warfare. Although biological weapons were not used in World War I, they were banned along with chemical weapons by the Geneva Protocol of 1925. Prompted by the poison gas attacks during the war, nations agreed in the Protocol to "the prohibition of the use in war of asphyxiating, poisonous or other gases, and of bacteriological methods of warfare."

Although the terms of the Protocol have been violated occasionally, the use of chemical or biological weapons was largely seen as aberrational and discredited. (Some worry that this may be changing in view of the failure of the international community to condemn Iraq vigorously for its repeated chemical attacks against Iran in the mid-1980s.)

During World War II, the United States developed extensive biological and chemical arsenals, and maintained them for years afterward. In 1969, however, at the suggestion of scientific and foreign policy advisers, President Richard Nixon announced that the United States would destroy its biological weapons and renounce biological warfare entirely. The use of biological weapons, he said, is "repugnant to the conscience of mankind" and could produce "massive, unpredictable and potentially uncontrollable consequences."[1] This unilateral decision led to the 1972 Biological Weapons Convention, signed by the United States, the Soviet Union, and more than 100 other nations.

Parties to the 1972 Convention agree "never in any circumstances to develop, produce, stockpile or otherwise acquire or retain . . . biological agents or toxins . . . of types and in quantities that have no justification for prophylactic, protective or other peaceful purposes." This phrase has been interpreted to mean that defensive research is permitted, though throughout the 1970s U.S. defensive research was minimal.

During the 1980s, however, the annual budget for biological defense research increased considerably. From $15 million in 1981 it climbed to $90 million in 1986, and has remained at about $60 million in each of the two succeeding years.[2]

The rise was fueled in large measure by the U.S. Administration's determination that the Soviet Union was violating the terms of the 1925 Geneva Protocol and the 1972 Biological Weapons Convention. The government's position was based on three allegations: that a 1979 epidemic of anthrax in Sverdlovsk was the result of an accident at an illegal biological warfare facility; that the Soviets or their surrogates in Southeast Asia and Afghanistan were using "yellow rain" toxins; and that the Soviet Union was secretly trying to create novel pathogens to be used as weapons. Several U.S. scientists and scholars challenged the validity of the government's accusations, and evidence has emerged that has placed the government's contentions in doubt. Although the Administration has never formally retreated from its position, in recent years it has muted its accusations. Nevertheless, many of the research projects and plans to expand biological defense facilities continue to move forward.

The activities of U.S. scientists in relation to biological warfare issues may be divided into three categories. The first includes those who have signed petitions or organized in opposition to aspects of the biological defense program. The second encompasses the scientists who work in the program either in army laboratories or under government contract in nonmilitary institutions. The third includes scientists who independently tried to test the government's allegations about the Soviets, and who were able to develop evidence that challenged the credibility of the government's case. Each category offers a distinctive ethical perspective about current biological warfare issues.

PETITIONERS AND ORGANIZERS

No group of scientists has been more persistent in its challenge to the government on the issue of biological warfare activities than the Boston-based Committee for Responsible Genetics (CRG). Established in 1983 amid concern about the implications of biotechnology on society, the CRG includes dozens of eminent biological scientists on its 60-member advisory board.

The organization's techniques for challenging and probing on the matter of the Army's research program have been quite traditional. It has publicized its concerns through newsletters, it has had spokesmen testify at congressional and Army hearings, and it has solicited signatures of medical and biological scientists to nationally circulated petitions.

A 1985 petition expressed concern that the Army's interest in "learning how to enhance the pathogenicity of microorganisms" could lead to "the eventual production of new biological weapons." Proclaiming opposition to the use of biology for military purposes and urging prohibition of secret military biological research, the petition called on "scientists throughout the world not to participate in research associated with the development and production of biological and chemical weapons." The document drew about 4,000 signatures from scientists and others, and was circulated among congressional and other public officials. More recently, another CRG-sponsored petition called on biologists and chemists to pledge "not to engage knowingly in research and teaching that will further the development of chemical and biological warfare agents." By the end of 1988, about 600 scientists had signed.

Other individuals and groups, conspicuously Jeremy Rifkin and his Foundation on Economic Trends, have through court challenges opposed the Army's plans to expand its biological research facilities. The CRG, however, has been the principal purveyor of the message among scientists that their work in the military program might be inappropriate and unethical.

Although the effectiveness of the CRG's efforts is difficult to measure, the organization has no doubt helped inform an audience of scientists and others about current military–biological activities. Many who signed the petitions might not otherwise have been stimulated to consider their own possible roles in military-sponsored research.

In 1988, several Utah scientists independently organized to influence the outcome of a biological defense project in their state. The Army had previously announced its intention to build an aerosol test chamber at Dugway Proving Ground, about 70 miles from Salt Lake City. The chamber was to achieve the highest level of containment (that is, biosafety level 4, or BL4), which could allow experimentation with novel pathogens that cause disease for which there is no cure. A group of University of Utah biologists and medical doctors obtained the signatures of more than 140 of their peers in opposition to the military's research plans.

The petition's coordinator, biology professor Naomi Franklin, noted that petitioners included five department chairmen and 85% of the biology department at the university. They expressed concern that the research plans seemed "flawed, hazardous and likely to break constraints of the 1972 Convention."[3] Several of the scientists met with their political representatives. In the process, they helped develop among these officials, the public, and the media a powerful opposition to the Army's plans to build its BL4 facility at Dugway. The Army altered its plans as a consequence, and announced that it would build a lower level, BL3, containment chamber there. In practical terms, the results of the scientists' efforts remain unclear. The testing of novel pathogens will evidently not take place at Dugway, but the Army has access to BL4 laboratories elsewhere.

Although they did not use the term "ethics" in their petitions, the scientists who joined the CRG and the Utah efforts based their actions on ethical premises. They viewed the Army's activities as inappropriate and implicitly unethical. Their efforts to lobby and affect policy reflected long-standing techniques of expression in the American political system. They educated a large audience, influenced decision makers, and prompted policy change.

SCIENTISTS WITH MILITARY FUNDING

The second category includes scientists whose work is funded by the Department of Defense. In the matter of biological warfare research, these scientists are particularly interesting, because their work is supposed to be limited by international treaty and domestic laws.

No doubt some who work directly for the Army have weighed the ethics of their assignments. In a recent letter to *Science* magazine, an investigator at Fort Detrick, Maryland, the Army's biological defense headquarters, expressed resentment at being considered a "biowarrior." Her research was for defensive purposes only, she wrote, and might even offer benefits to the civilian population. Although she did not discuss the possible military applications of her work, the scientist did show concern about how it was perceived.[4]

Others in the program, however, feel that they are simply doing a job. This was made clear to me during interviews in 1983 and 1984 with the public affairs officer at Fort Detrick. Speaking with pride about scientists in the Army's program there, he said, "They are dedicated to science, not politics." Their attitude, he said, is "Just leave me to my work, and I'll produce for you." If the public affairs officer is correct, ethical questions peculiar to their work were not among their principal concerns.[5]

In addition to these scientists are independent investigators who, after receiving grants from the Army, conduct research at universities or other nonmilitary institutions. These investigators usually write proposals within broad areas known to be of interest to the Army. For example, in 1985 the U.S. Army Medical Research and Development Command at Fort Detrick advertised in *Science* magazine for proposals for "research in staphylococcal toxins of military importance." The advertisement suggested "studies involving specific mechanisms that provoke symptoms following oral, intravenous, and pulmonary routes of exposure to toxins," and "recombinant DNA studies that may lead to new immunogens or antigens for use in diagnostic or immunization procedures."[6] Presumably such research could be scientifically challenging and result in information that would enhance the health of the general population. By 1988 more than 100 Army-funded research projects were underway.

Unclear, however, is whether the scientists who won these contracts considered issues beyond the scientific and technical aspects of their proposed projects. Two areas in particular relate to larger questions about their research that many are not likely to have thought about — the purpose the military intends for their findings, and the relationship of their work to international and domestic laws.

In recent years, many have wrestled with questions about the responsibility that a scientist bears for the effects of his research if they turn out to be harmful. Some have suggested that even scientists who perform basic research (that is, research not necessarily having any intended application) do not escape responsibility. The "imagination principle," according to Daniel Callahan of the Hastings Center, obliges a scientist to conjecture how his research might lead to harmful consequences, even in improbable circumstances. Only after examining alternative scenarios with the care that he devotes to considering alternative technical approaches to an experiment, can the scientist fairly make a moral judgment about the nature of his work.[7]

All the more is this true in the area of applied research where ends are envisioned. This does not mean that because a scientist may imagine a dangerous scenario that his work is necessarily immoral and ought not be pursued. But his calculations should include considerations of the likelihood of danger, its extent, and whether any good might result that would outweigh the bad. Such an assessment, when seriously undertaken, places the scientist in a more ethically sound position to decide whether to pursue a project or not.

In consideration of biological warfare issues, ethical judgments are augmented by legal strictures that are more proscriptive than in other areas of military research. Any scientist under contract with the military should be familiar, for ethical as well as legal reasons, with applicable statutes and regulations. In the area of biological research, some work might be inappropriate while not being explicitly illegal.

Differentiating between offensive and defensive biological research is extremely difficult. Both forms of research might require dealing with pathogens that are potential biological weapons. The difference is commonly viewed as depending on the quantity of materials being used or developed, and the intent of the user. But these criteria are quite subjective. How many pathogens in how many laboratories are too many to be considered necessary for defensive research?

The matter would be clear if a scientist were asked to grow warehousefuls of secretly developed pathogens. But the issue is not likely to be that simple. In any case, research that is considered legal may still have important ethical implications. For example, a $250,000 contract was granted in 1985 to a scientist at Brigham Young University to develop a vaccine against the bacterium that causes anthrax. *Bacillus anthracis* is a potential biological weapon, and the project was directed at removing the genetic component that is responsible for the disease. The attenuated organism would then be used to try to develop a vaccine. But let me pose another conceivable use of attenuated *Bacillus anthracis*.

For many years, the Army has conducted field tests with so-called simulants of biological weapons, that is, bacteria that are less harmful than those that might be used as biological weapons. These simulants, usually *Serratia marcescens* and *Bacillus subtilis*, were secretly sprayed over populated areas throughout the United States as part of an Army research program. Millions of people were kept in total ignorance as they were exposed and their health put at risk (the Army never monitored the health of the exposed populations). The Army apparently is not spraying in heavily populated areas at the moment, though it reserves the right to do so in future "vulnerability" tests. In recent reports, it has expressed hopes of finding better simulants, that is, ones that supposedly are harmless yet more closely resemble actual biological warfare agents.

If the Army could attain an organism only a gene away from being a real warfare agent, this might well be considered the ideal simulant it has been searching for. Viewed as harmless, such organisms might be sprayed into the New York City subway system, and over San Francisco and other cities, as was done with less ideal simulants in the past, to assess their survivability and dispersion patterns. As in the past, Army scientists might again ignore biological and health specialists who hold that spraying *any* so-called simulants at unsuspecting people is unethical and risky to their health.[8]

Did the scientist working to attenuate the anthrax bacillus consider this scenario? Did he even know about the open air testing program? If he did, how much weight did the knowledge have on his decision to bid for the project grant?

Questions like these represent the core of the dilemma for a biological scientist who considers the practical consequences of his research for the military.

THE OMBUDSMEN

Scientists in the third group of those acting in relation to biological warfare issues have been quite influential despite their small numbers. Prompted by skepticism about U.S. claims concerning illegal Soviet activities, these scientists systematically investigated the claims and found them wanting. The government's charge about "yellow rain" was the only allegation based on supposedly direct scientific evidence. It received the most public attention and offers vivid instruction.

Late in 1981, spokesmen for the Departments of State and Defense declared that planes had released clouds described as yellow rain over defenseless people in Laos, Kampuchea, and Afghanistan. The people allegedly became ill and died. U.S. government investigators found that samples of yellow rain contained trichothecene toxins from fungi, and concluded that the Soviets were using the toxins as chemical or biological warfare agents. (Biological warfare agents are generally understood to be living microorganisms, but the Biological Weapons Convention also applies to toxins, the chemical products of organisms.) They assumed that the trichothecenes were not found naturally in the combinations identified in the samples collected in Southeast Asia.

Matthew Meselson, a molecular biologist at Harvard, believed that the government's contentions were speculative. He could find no studies that confirmed whether or not the mycotoxins occurred naturally in Southeast Asia, or that they caused massive hemorrhaging and death, as the United States claimed.

The government in subsequent years continued to affirm that the Soviets or their surrogates were "flagrantly" violating the chemical and biological agreements.[9,10] During this time, Meselson, and several other scientists who had remained unconvinced, began an odyssey in which they tested the government's claims. Besides having questions about the symptoms of the supposed victims, Meselson was impressed that separate investigations sponsored by Canada, Australia, and the United Nations had found pollen in the yellow rain samples. Discussions with specialists in botany and bees, and his own micrographic studies, led him in 1983 to suggest that yellow rain might be bee feces. The proposal was ridiculed by government spokesmen, especially by Sharon Watson, the government's original proponent of the toxin-weapon theory, who dismissed Meselson's notion as amusing.

The following year, Meselson received an award of several thousand dollars from the MacArthur Foundation, some of which he used to inquire further into the yellow rain question. Accompanied by Thomas Seeley, a bee expert at Yale, and Pongthep Akratanakul, a Cornell-trained Thai entomologist, he visited honeybee nesting locations in Thailand. The scientists found several areas of yellow-spotted vegetation, and they determined that the yellow spots were fecal deposits. At one point, they were caught in a fecal shower. Dozens of spots fell on them which, they reported, closely resembled "the showers and spots said to be caused by yellow rain."[11]

After this experience, the scientists visited a refugee camp where alleged victims of yellow rain attacks had previously given interviews that became part of the U.S. evidence. But now, as the three scientists showed several groups of refugees leaves spotted with bee feces, most refugees could not identify the spots for what they were—bee feces—and some thought they were chemical or biological warfare agents.[11]

Other scientists, including British chemical and biological warfare expert Julian Perry Robinson, joined inquiries into the yellow rain issue, and concluded that yellow rain was never anything more than bee feces. By 1985, the United States acknowledged that Soviet toxin weapons were no longer being used, but it continued to contend they had been used earlier. Although opportunities to confirm or refute the U.S. allegations have dimmed with time, Meselson and several colleagues are more convinced than ever of their case. Moreover, recently declassified documents show that some U.S. investigators had doubts all along about the validity of the government's position, but that these doubts had been kept from the public.[12]

One of the most compelling arguments that placed the Administration's contentions in doubt came from the on-the-ground experiences of Meselson and his associates in Southeast Asia. Had they not traveled to the area and investigated personally, their case would have been weaker, and the challenge to the government less convincing. Their approach suggests a means by which investigations of similar controversies might be formally institutionalized. In policy disputes based on allegedly scientific evidence, a scientific ombudsman, nominated by an independent body like the National Academy of Sciences, might be funded to investigate and challenge the government's findings. Although we do not explore here the possible variations of such an arrangement, the general principle deserves consideration.

In the yellow rain case, the actions of a dozen or so scientists were fortuitous. These actions were largely inspired by Meselson's singular dedication to the issue, which was augmented by unexpected private funding.

Scientific disputes are usually conducted within the scientific community—in publications and conferences. With public policy issues like yellow rain, however, these avenues to the "truth" might be more restricted. On policy matters, particularly those relating to the military, the government may control information released to the public. One cannot be sure that all pertinent material—and doubts—have been laid out. Furthermore, insofar as government agencies often have an interest in promoting a particular policy, their interests may skew their interpretation of ostensibly scientific findings. The Reagan Administration used the yellow rain charge to help vindicate its suspicions about Soviet cheating on international agreements. Supposedly scientific findings were announced in support of allegations it was predisposed to believe, and doubts among government scientists about the yellow rain claims were never publicly acknowledged.

The Administration's stand on yellow rain was in line with other charges about alleged Soviet violations. After reviewing claims that the Soviets were developing novel bioweapons, several respected scientists found the government's conclusions questionable. Similarly, doubts about Administration charges concerning the 1979 anthrax episode in Sverdlovsk, which the Soviets blamed on the distribution of anthrax-infected meat, were reinforced by later disclosures.[13] Yet these cases in large measure were responsible for the vastly expanded U.S. biological-military program in the 1980s. They stand as warnings about how the authority of science can be used with doubtful warrant for political purposes.

CONCLUSION

In sum, many scientists have shown concern about the direction and ethics of the U.S. biological defense program. The lessons of current biological warfare issues have implications for other areas of ethics and military research as well. Based on the overview presented in this paper, several policy approaches might be considered:

1. The military authority that sponsors a research project should make available pertinent laws, regulations, and international agreements to the scientists and others involved.

2. Scientists engaged in military projects should have an ethical responsibility to know and understand the laws, regulations, and agreements associated with their work.
3. An independent consultative agency might be established, perhaps on the order of the Institutional Review Boards that now assess experiments dealing with human subjects, to review questions of ethics and biological warfare research.
4. Through the consultative agency or other independent means, a procedure should be established to assess scientific claims whose ethical or technical validity are under question. Variations on previously suggested ideas—a science court, for example—might be appropriate.
5. The desirability of recruiting and funding scientific "skeptics" should be considered. This could be done in the form of scientific ombudsmen, nominated perhaps by the National Academy of Sciences, who would test the scientific validity of a claim.

I present these ideas as discussion points, not refined proposals for action. The needs, however, are clear. As biological warfare issues in the 1980s have demonstrated, government policies and large-scale expenditures can be based on questionable (if nominally scientific) evidence. In the end, a heightened ethical awareness among scientists, including consideration of the possible consequences of their research, would be in the best interest of science, and the nation.

REFERENCES

1. White House Press Release. 1969. Remarks of the president on announcing the chemical and biological defense policies and programs. November 25.
2. ANON. 1987. Funding in the biological defense program. Genewatch 4(4-5): 3.
3. DAVIDSON, L. 1988. U. experts condemn germ lab. Deseret News (Salt Lake City). August 23: B1.
4. NOVAK, J. M. 1988. Research at USAMRIID (letter). Science 242(4876): 168.
5. COLE, L. A. 1988. Clouds of Secrecy: The Army's Germ Warfare Tests over Populated Areas: 38. Rowman & Littlefield. Totowa, NJ.
6. U.S. Army Medical Research and Development Command. 1985. Invitation to submit proposals for research in staphylococcal toxins of military importance. Science 227(4686): 468.
7. CALLAHAN, D. 1976. Ethical responsibility in science in the face of uncertain consequences. Ann. N.Y. Acad. Sci. 265: 1-12.
8. Subcommittee on Health and Scientific Research, Committee on Human Resources, U.S. Congress. 1977. Biological Testing Involving Human Subjects by the Department of Defense (Senate hearings on March 8 and March 23): 261-296. U.S. Government Printing Office. Washington, DC.
9. U.S. Department of State. 1982. Chemical Warfare in Southeast Asia and Afghanistan: Report to the Congress from Secretary of State Alexander M. Haig, Jr., March 22, 1982. Special Report No. 98. N. Howard & C. Sussman, Eds. U.S. Department of State, Bureau of Public Affairs, Office of Public Communication, Editorial Division. Washington, DC.

10. TAUBMAN, P. 1982. U.S. offers report to show Soviet use of chemical war. New York Times. March 23: A1.
11. SEELEY, T. D., J. W. NOWICKE, M. MESELSON, J. GUILLEMIN & P. AKRATANAKUL. 1985. Yellow rain. Sci. Am. **252**(3): 137.
12. ROBINSON, J., J. GUILLEMIN & M. MESELSON. 1987. Yellow rain: The story collapses. Foreign Policy. No. 68: 100–117.
13. COLE, L. A. 1988. Clouds of Secrecy: chapters 9–10.

Ethical Elements of Chemical/Biological Defense Research

SHIRLEY A. LIEBMAN

Geo-Centers, Inc.
Fort Washington, Maryland 20744

INTRODUCTION

Mission-oriented research in industrial and government laboratories requires 1) individual acknowledgment of corporate or research directorate goals and 2) the personal commitment to achieve them. As a research contractor at a federal laboratory, I view these requirements to mean there is full awareness and understanding of the design and conduct of the contracted research program to meet these goals. It is evident to me that dangers persist both in our domestic and international scenes that call for defense research at a necessary and sufficient level.[1-5] To define what is necessary and sufficient research, one must resort to subjective and often controversial evaluations. These evaluations are difficult, even for the most experienced practitioners in the field. When contractors work with informed, knowledgeable government research managers, efforts are directed to the defined needs within the competitive contracting process.

In the conduct of government-funded research, there are many parallels with research conducted in the industrial scene. Sensitive, proprietary research is important to many competitive industries, as well as to military-related work. Investigators are often required to maintain confidentiality where it is deemed appropriate by the organization. The level of required security is best determined by the technical, legal, and senior management staff, which bears responsibility for the corporate or national well-being. Perhaps this point is least convincing to some academics when they weigh national security concerns against what they perceive as their scientific rights. In my view, it is the direct responsibility of an employee or funded researcher to accept, negotiate, or reject the contracted limits of research proposals. This is one potential conflict of individual versus corporate/military rights that most pointedly brings out the personal attitudes and professional ethics of the researcher.

ETHICAL PERSPECTIVES

Three considerations support my personal commitment to the conduct of the contracted research:

1. *The Constitutional Basis*, wherein support of our government to provide for the common defense represents a fundamental and moral responsibility of all citizens.

2. *The Ethical Concerns*, wherein personal beliefs in religious and moral standards are acknowledged as being consistent with our rights to life, liberty, and the pursuit of happiness.

3. *A Personal Attitude*, wherein contributions are made with a particular mind-set to protect these collective and individual rights.

The Constitutional Basis

Threats to the health and safety of our domestic and international communities continue to make us aware of the need to counter violent, destructive, and terrorist elements. Defense-oriented efforts are required to identify those responsible for these elements and to deter aggression. Moral dictates held by citizens result in a range of actions consistent with such efforts.

The Ethical Concerns

Quiet discussions with friends and neighbors who belong to religious groups that forbid military involvement for their members are a source of inspiration and moral support for my activities involving government/military-based research. Because their belief system prohibits them from military service, they proudly cite their contributions carried out during wartime under conscientious objector status. Their full recognition that "somebody has to do it" is consistent with their understanding that real dangers persist throughout our society, requiring the existence of our local police and the need for our government to provide for the common defense. Thus, we carry out our individual actions with mutual respect, according to our own moral and ethical judgments and in line with our knowledge about the existing or potential danger.

I recognize many in my own congregation who profess similar pacifist beliefs, but who do not deny that there are threatening situations that must be met with determined actions. Our liberal religious outlook possesses sincere spiritual and humanistic values seen from many perspectives and encourages a way of life consistent with them, as well as an open attitude toward dissent.

Dissenting opinions have an important role in our society: they serve to highlight and to focus attention on important issues. Dissent often presents an opportunity for discussing alternatives and presenting critical evaluations. The intent and individual mind-set of those involved should be considered in the evaluation of such dissent.

A Personal Attitude

Research in the field of analytical chemistry is entirely in agreement with my ethical code: this research benefits the public by solving medical, environmental, industrial, and defense problems. Besides the general obligation to benefit the public, there are several principles that guide research in my profession: acceptance of corporate and government codes of ethics; observance of appropriate safety and information security protocols; and accountability for the quality of one's own work, which includes making technically honest, "best judgment" decisions.

Many professional groups, including the American Chemical Society and the American Association for the Advancement of Science, support educating the public on scientific and technological issues. These groups, through their meetings and publications, perform the valuable service of improving the quality of public discussion in the areas of technology, science policy, and risk communications. Fair and intelligent decisions are more likely if the public is informed in an accurate, timely manner.

The need for an informed public is particularly important in the area of risk assessment. Decisions bearing on chemical, biochemical, and military research depend to a great extent on how risks are assessed. Public attitudes toward chemical and biotechnology industries, as well as military research, show the need for enlightenment in the risk assessment area.

RISK ASSESSMENT: PERCEIVED AND REAL THREATS

A good illustration of an appropriate response from an informed public using the democratic process was the instance in which a vocal minority brought their goal, the shutdown of two nuclear energy plants, to a state referendum in 1988. A two-to-one defeat of that referendum would seem to indicate that the reality of the need for safe, efficient nuclear power is readily understood by the majority of our citizens. Similar controversies in risk assessment dealing with biotechnology, foods, hazardous wastes, and the environment will, one hopes, be resolved with realization of the technical complexities and the pragmatic, knowledgeable efforts required to answer both short- and long-term needs.

It should be evident that all chemical and biological research has inherent risks. Safety procedures, however, are an integral part of training for anyone dealing with hazardous materials. The chemical industry has embedded safety measures into every facet of its operation. We always strive to minimize the risk element, along with the cost and time factors consistent with a rigorous technical approach. I have felt much more at risk driving 30–35,000 miles per year on our highways than I have felt conducting my industrial and military research over a period of nearly thirty years.

It is a foregone conclusion that we should do our part to stop abuses and incompetence that increase risks in our society. This is true whether the abuses occur in our technical, judicial, military, social welfare, academic, or business activities. Self-governance within our own professions is mandatory if energies are to be spent in productive directions.

Those who would appear to be qualified to offer their opinions do our society a disservice, however, when they mount unjustified attacks on environmental, industrial, or military research programs. Such attacks tend to distort issues that are, of themselves, technically difficult and that involve controversial measures to solve them. Rather than helping to solve the problem, some critics speaking outside their areas of expertise become part of the problem. On the other hand, valid criticism of poorly designed or conducted programs is valuable, so that the inadequate research may be identified, improved, or eliminated. Again, the motivation and the knowledge of those who challenge the quality of research are important considerations in assessing the value of the dissent. Strategic planning of major research programs requires complex evaluations by qualified senior management to ensure that the science and technology base is adequate to meet corporate/national needs. Decision making in this area is not simple.

Personal mind-set also plays a role as we meet our ordinary and not-so-ordinary risks and challenges. My family's attitude is reflected in the theme of Theodore Roosevelt's "Dare Greatly": "It is not the critic who counts, nor the man who points out how the strong man stumbled, or where the doer of deeds could have done better. The credit belongs to the man who is actually in the arena . . ." This theme states our attitude toward dealing with local and national problems, even when we are critical of some events around us and are aware of real risks. To be sure, we do not live in a risk-free environment, but we try to cope with risks in a constructive manner. When this viewpoint is turned toward positive directions, it is seen that technical innovation and economic risk taking is the stuff entrepreneurs thrive on.

This personal and professional ethical perspective thus forms my background for approaching the issue of chemical/biological defense research and underlies my commitment, made with a clear conscience, to the contracted defense effort.

FOCUSED DEFENSE RESEARCH: NEED AND RESPONSE

The task with which I am associated is to detect and to identify dangerous chemical and biological agents that may be present in the environment.

This task, which is a team effort, requires interdisciplinary knowledge of researchers for multimedia environmental sampling (air, water, soils), specialty instrumentation used in advanced analytical chemistry, data analysis expertise from chemometricians, and computer-assisted interpretive guides developed using applied artificial intelligence tools.

The research entails developing and using the most advanced equipment and methods in chromatography, spectroscopy, thermal analysis, computer science, and allied fields. These efforts, which are conducted within the general field of trace organic analysis, yield results that are directly relevant to the materials, environmental, and forensic sciences. Interpretations of these studies influence the risk assessment methods that are used to determine current medical and environmental civil defense requirements.[5]

Results from relevant research with my government colleagues are given in the references.[6-16] These published contributions, made under the National Research Council Senior Associate Program and on-site contract research at a federal laboratory, have passed the traditional peer-review system and have been presented at national and international scientific meetings. Such research is open to critical comments and represents the best of our knowledge in a clearly defined manner and within the limits of the experimental designs used in studies of complex chemical systems. For example, our applied basic research into chemical mechanisms[8,9] parallels work on fundamental decomposition mechanisms for which the 1987 Nobel Prize in chemistry was awarded to investigators at the University of California, Berkeley.[17]

In general, technical exchanges are sought and advice or critical comments from qualified persons are welcome as we combine our problem-solving capabilities to achieve the needed goals. Personal communication via informal networks with professional colleagues, meeting attendance, or other interactions throughout our scientific community all help to maintain a high quality of research.

On the other hand, self-styled experts with rudimentary knowledge in these and other technical areas should exercise restraint when offering comments, particularly if these comments are primarily made for self-serving purposes and in a disruptive or contrary manner. Indeed, persons with limited understanding should defer critiques of nuclear engineering technology and evaluations of reactor safety management or toxic materials handling to those knowledgeable in the field. Clearly, uninformed persons cannot fulfill the role of advising our research directors or science policy makers if we are to meet the actual challenges before us.

TECHNOLOGY ASSESSMENT, FORECASTING, AND TRANSFER

In the competitive industrial scene, the development of corporate strategy relies on technology assessment methods. Technical "gatekeepers" help in identifying new directions for research, in monitoring the actions of competitors, and in advancing the research and development planning process. Technology assessment and forecasting informs not only corporate management, but also those responsible for government and military research planning.[18-22]

The role of technology transfer[21] is significant. The transfer process leads to expedient, effective use of the generic technologies that are developed in many government research studies. A specific illustration is related to supercritical fluid technology, initially used in our military research for the safe analysis of propellants. Commercial applications of instrumentation developed in our research are now used throughout food, pharmaceutical, chemical, petrochemical, and biomedical laboratories. Furthermore, the design, production, and use of high-performance materials in our automotive and aerospace industries are additional evidences of the importance of the research infrastructure to our economic well-being and the defense of our nation. Gansler has discussed the need to integrate our civilian and military industries more effectively.[18]

Industry/government interactions can help bridge the gap between our civilian economy and military-related programs. Technology assessment, forecasting, and transfer are essential in accomplishing this task. The methods of our typical industrial research team have proved to be highly successful in such mission-oriented and applied basic research efforts. In this manner, research contractors work with government colleagues to achieve a more productive, responsive research and development effort. The managers of the Small Business Innovative Research Program are also important in accomplishing these contracted goals.

When he accepted his 1987 Scientist of the Year Award, Ralph E. Gomory, Vice-President of Science and Technology at IBM, emphasized that "predominant science does not mean predominant product" and that understanding the manufacturing/development process is the key to industrial leadership.[23] Our innovative and productive industrial teams must improve the sequence from research and development to manufacturing and delivery of "the quality product." Helmut Hellig of the National Institute of Standards and Technology adds that traditional roles must change for processing, quality assurance, and technical development.[24] The chemical industry contributes greatly to our economic and defense posture, and positive actions in these areas have been discussed.[25]

The source of my motivation is to participate in these dynamic industry/government research and development programs with colleagues working at leading-edge technologies.

SUMMARY

Personal and professional ethics both shape attitudes toward scientific research programs. Personal ethics are presented from the viewpoint of shared community responsibility and individual integrity combined with beliefs that guide relevant responses to defined needs. Professional ethics include personal motivation, accountability, and technical honesty as basic requirements for all research and development efforts, including defense-oriented chemical and biological research. Industrial and military research issues of safety, confidentiality, and technical quality are addressed. Specific documented research in the field of trace organic chemical analysis is cited to show how federal laboratory programs contribute to the general scientific community. Technology assessment and transfer are considered as important means to ensure productive research efforts leading to economic benefits.

ACKNOWLEDGMENTS

I appreciate the insight and patience afforded me by my government and industrial colleagues.

REFERENCES

1. ROBERTS, B. 1986. Chemical weapons: A policy overview. Issues Sci. Technol. Spring: 102-114.
2. SHAPIRO, J., J. DIEBOLD, S. GORTON & W. MASSEY. 1986. A national defense policy. Issues Sci. Technol. Spring: 116-125.
3. PEARSON, G. S. et al. 1988. Chemical defense. Chem. Br. **24**(7): 657-691.
4. HENDRICKS, M. 1988. Germ wars. Sci. News **134**: 392-395.
5. MERSHON, M. M. & A.V. TENNYSON. 1987. Chemical hazards and chemical warfare. J. Am. Vet. Med. Assoc. **190**: 734-745.
6. LIEBMAN, S. A., P. M. DUFF, M. A. SCHROEDER, K. D. FICKIE & R. A. FIFER. 1986. Degradation profile of propellant systems with analytical pyrolysis/concentrator/GC technology. J. Hazard. Mater. **13**: 51-62.
7. LIEBMAN, S. A., P. M. DUFF, M. A. SCHROEDER, R. A. FIFER & A. M. HARPER. 1986. Concerted organic analysis of materials and expert system development. In Artificial Intelligence Applications in Chemistry. American Chemical Society Symposium Series, No. 306. T. H. Pierce & B. A. Hohne, Eds.: 365-384. American Chemical Society. Washington, DC.
8. LIEBMAN, S. A., A. P. SNYDER, J. H. KREMER, D. J. REUTTER, M. A. SCHROEDER & R. A. FIFER. 1987. Time-resolved analytical pyrolysis studies of nitramine decomposition with a triple quadrupole mass spectrometer system. J. Anal. Appl. Pyrolysis **12**: 83-95.
9. SNYDER, A. P., J. H. KREMER, S. A. LIEBMAN, M. A. SCHROEDER & R. A. FIFER. 1989. Characterization of RDX decomposition by N,H isotope analyses with pyrolysis atmospheric pressure ionization tandem mass spectrometry. Org. Mass Spectrom. **24**: 15-21.
10. HARPER, A. M. & S. A. LIEBMAN. 1985. Intelligent instrumentation (proceedings of a chemometrics workshop, May 1985). J. Res. Natl. Bur. Stand. Spec. Publ. **90**(6): 453-464.
11. LIEBMAN, S. A. 1988. Supercritical Fluid Applications in Materials Science. Paper presented at the 27th Annual Conference on the Practice of Chromatography (Committee E-19 on Chromatography). October 10-12. Baltimore, MD. American Society for Testing Materials. Philadelphia, PA.
12. LIEBMAN, S. A., R. R. SMARDZEWSKI, E. W. SARVER, D. J. REUTTER, A. P. SNYDER, A. M. HARPER, E. J. LEVY, S. LURCOTT & S. O'NEILL. 1988. Analytical Instrumentation and Applied Artificial Intelligence in Materials Science. Proceed-

ings of the ACS Division of Polymeric Materials: Science and Engineering. Vol. 59: 621-628. American Chemical Society. Washington, DC.
13. LIEBMAN, S. A., E. J. LEVY, S. LURCOTT, S. O'NEILL, J. GUTHRIE & S. YOCKLOVICH. 1989. Integrated intelligent instruments: Supercritical fluid extraction, desorption, reaction and chromatography. J. Chromatogr. Sci. **27**: 118-126.
14. SNYDER, A. P., C. CROSS, S. A. LIEBMAN, G. A. EICEMAN & R. A. YOST. 1989. Thermal desorption–Tandem mass spectrometry for detection and identification of formulated products. J. Anal. Appl. Pyrolysis **16**(3): 191-204.
15. LIEBMAN, S. A. & A. P. SNYDER. 1989. An overview: Analytical toxicology and integrated intelligent instruments (I^3). *In* New Technologies and Concepts for Reducing Drug Toxicities. In review. Telford Press. Caldwell, NJ.
16. WINDIG, W., S. A. LIEBMAN, M. B. WASSERMAN & A. P. SNYDER. 1988. Fast self-modelling curve-resolution method for time-resolved mass spectral data. Anal. Chem. **60**: 1503-1510.
17. ZHAO, X., E. J. HINTSA & Y. T. LEE. 1988. Infrared multiphoton decomposition of RDX in a molecular beam. J. Chem. Phys. **88**(2): 801-810.
18. GANSLER, J. S. 1988. Integrating our civilian and military industry. Issues Sci. Technol. Fall: 68-73.
19. U.S. Office of Technology Assessment. 1988. Defense Technology Base: Introduction and Overview. U.S. Government Printing Office. Washington, DC.
20. JONES, H. & B. TWISS. 1978. Forecasting Techniques for Planning Decisions. Petrocelli. New York, NY.
21. World Future Society. 1989. Government with Foresight: Successes and Challenges in Preparing for the 21st Century (seminar). Washington, DC.
22. ALLEN, J. T. 1978. Managing the Flow of Technology. MIT Press. Cambridge, MA.
23. GOMORY, R. E. 1987. Dominant science does not mean dominant product. Res. Dev. November: 72.
24. HELLIG, H. 1988. Technology and the competitive challenge. Res. Dev. July: 44.
25. GOOD, M. L., Ed. 1988. Biotechnology and Materials Science: Chemistry for the Future. American Chemical Society. Washington, DC.

The Place of Department of Defense-Sponsored Research at the University[a]

JOSEPH P. MARTINO

Research Institute
University of Dayton
Dayton, Ohio 45469

The title, which I have worded with special care, emphasizes the first of several important distinctions that I will make.

Officials of the Department of Defense (DOD), staffs at their laboratories, and engineers at defense contractors, are perfectly capable of reading the scientific literature. Research sponsored by the National Science Foundation and the National Institutes of Health not only can be but has been used by the DOD. Neither the researcher nor the researcher's institution can foresee whether a particular piece of research will find military applications. The event over which the researcher and the researcher's institution do have control is acceptance of research sponsorship by the DOD. Therefore, the issue which I wish to address is the ethics of accepting such sponsorship.

Before presenting my analysis of DOD-sponsored research in universities, however, I will include, for the sake of providing a full disclosure, some statements about my own institution.

The University of Dayton Research Institute is entirely supported by external grants and contracts. It does not draw upon the University's financial resources, but instead supplements them by paying for the use of the University's facilities. In the last fiscal year, the Institute had an income of $30 million. These funds supported about 450 full-time staff members, including 250 professional researchers. In addition, they provided partial support to about 60 faculty members, nearly 200 undergraduate students, and nearly 100 graduate students. Thus, although the Institute has its own full-time professional staff, it is also integrated with the University's academic staff.

Over two-thirds of these funds came from a single agency, the U.S. Air Force. No other single source provided more than 10% of the total funding. Thus, the dominant source of our funds is not even the entire DOD but only a part of it.

---

[a] The opinions presented in this paper are those of the author, and do not necessarily represent the positions of either the University of Dayton or its sponsors.

This dominance of our funding by the DOD (or a part of it) is a matter of concern on the campus, especially because ours is a university with a Catholic tradition. Some faculty and students object strongly to the Institute's involvement with the DOD. We in the Institute are of course aware of the issues involved. It is unlikely that there are any arguments against DOD support that we have not already heard from our own colleagues, and that we have not already dealt with.

What are these arguments? I believe the various objections to DOD support of university research can be summarized as follows (Rewak[1] presents most of these):

1. All war is immoral; a university should have nothing to do with military activity.
2. Although defensive wars can be moral, weapons of mass destruction, such as nuclear weapons, are morally inadmissible, and a university should not conduct any research that might contribute to their development.
3. A university is devoted to the preservation and enhancement of culture. It should not support warlike activities because their practical result is the obliteration of culture.
4. Classified DOD research, which contradicts the demands of academic freedom, should not be conducted on a campus.
5. Although some amount of DOD research is acceptable, the current level of dependence of universities on DOD funding leads to militarization of the university.

I will now consider each of these objections in turn.

The argument that war is inherently immoral is pure pacifism. It denies the legitimacy not only of DOD funding of university research, but of the DOD itself. I reject this argument. I stand firmly in the Catholic tradition of the Just War Doctrine. Nonresistance is legitimate for some individuals; it is not legitimate for nations. Defense against unjust aggression is not only legitimate but obligatory for nations, and is a serious duty of public officials.

I also reject the argument that some weapons, such as nuclear weapons, are intrinsically immoral. I have argued this point at greater length elsewhere.[2] Here I will simply note that weapons are inanimate objects, totally without volition. They are incapable of rising even to the dignity of sinning. Only the weapon user is capable of being either moral or immoral. I insist that while nuclear weapons are easier to misuse than are gunpowder weapons, they can be used morally. The immorality in any specific case originates in the weapon wielder's misuse of the weapon, not in the weapon itself.

The third argument against DOD funding, that war subverts the very purpose of a university, is more subtle. Nevertheless, in my view it is still wrong. It accepts the legitimacy of a national defense, but insists that such a defense be conducted by others. It argues that the university must not dirty its hands in such a defense. The roots of this argument go back to the medieval papal prohibitions against the clergy bearing arms. Because all medieval universities were founded under religious auspices, for the formation and education of priests and professed religious, the prohibition against bearing arms was extended to university faculty and students. It is ironic in the extreme to hear this argument made today by faculty members of totally secularized universities.

Beyond the impropriety of claiming for a secular university a position originally established on religious grounds, there is the issue of the effects of unjust aggression on culture. We ought to keep in mind the statement attributed to Herman Goering: "Whenever I hear the word 'culture,' I reach for my pistol." Culture is obliterated not just by the physical destruction that comes with war. Culture is obliterated by the victory of tyrants. War may destroy the physical manifestations of culture. Tyranny stifles the minds which can produce and reproduce culture. If unjust aggression were allowed to triumph, more culture could be obliterated than if the aggression were to be defended against.

At a yet deeper level, there is the question of the place of the university in society. By joining a university, one does not abandon one's citizenship. Faculty, staff, and students still retain the obligations of citizenship, particularly the obligation to defend the society that supports the university. In carrying out this obligation, however, the members of the university community should not undermine the purpose of the university, for that would be to weaken their own society in another way. I'll return to this point later. Here I want to stress that those in the university cannot be indifferent toward aggressive threats from outside their society. When their society is threatened, they have an obligation to defend it.

The next argument, regarding classified research, does have some merit. The university is in the business of producing and disseminating knowledge. Classifying the knowledge it produces cuts short the very possibility of dissemination. The results cannot be published, and cannot be discussed in the classroom. Here again, however, some distinctions are necessary. Some research actually does generate information to which access is restricted by classification. This is the kind of research usually associated with the phrase "classified research." Other research, however, though it may require access to classified information, and thus qualify as classified research, may generate only unclassified information.

The first category, research that generates classified information, is the most restrictive. Even so, such contracts are not simply "black holes" from which nothing escapes. Institutions with such contracts are permitted to identify the sponsor and the general nature of the work. Researchers with such contracts are usually permitted to publish information about processes and procedures. Usually only operational details are classified.

The second category, where access to classified information is required, is the more common type. Here there are very few constraints on publication. One example from my own institution is research on birdstrikes: collisions between birds and aircraft. I will say more about this research below. For the moment, I will simply note that our researchers needed to know the speeds and altitudes at which combat aircraft operated, in order to determine the forces generated by birds striking the aircraft canopy. Some of this information was classified. The research performed, and the results produced, however, were completely unclassified.

About 10% of our research projects are classified. About 20% of these classified projects, or about 2% of our total work, generate classified information. Having carefully chosen the kinds of work we are willing to undertake, we find that we have very few problems with military classification. In fact, we find that industry places more constraints on us, to protect proprietary information, than does the DOD. Hence, even though there can be some legitimate concern over classifying university research, universities are free to determine for themselves the amount of government classified research, or industrial proprietary research, that they will undertake. They need not take on more classified or proprietary research than they consider consistent with their commitment to academic freedom. Within this limit, there is no merit to the objection to classified or proprietary research.[b]

The remaining objection to DOD funding is that of militarizing the university. This is part of a broader issue, that of militarizing society as a whole, or becoming a garrison state. Although we might submit to militarizing our society for a limited time during a great war, we would expect to return to a normal state at the end of the war. It would surely be a perversion of the very idea of defense if we militarized our society, in peacetime and for an indefinite duration, in the name of defending it against tyranny. Although this broader issue is certainly a matter for concern, I will focus on the narrower issue of militarizing the university.

Just as with our society, we might militarize our universities during a great war, but we would expect them to return to their normal functions at the end of the war. This was the case during World War II. Many universities, such as Harvard, turned over some of their classrooms to the military for officer training. Their faculty also trained military personnel in certain technical specialties. More pertinent to my topic, during World War II, universities such as the Massachusetts Institute of Technology diverted part of their research to the direct support of military weapons projects. The Radiation Laboratory at MIT is perhaps the most famous, but there were many other examples.[3] When the war ended, however, these universities returned to their normal peacetime activities.

One of the reasons we demilitarize our universities at the end of a war

---

[b] Some universities are finding that protecting their intellectual property through patents is even more constraining than either military classification or industry proprietary restrictions. Publication of research findings before a patent is granted can jeopardize a U.S. patent, and may be an absolute bar to foreign patents. Moreover, this constraint cannot be avoided simply by refusing to accept outside contracts from the DOD or from industry.

is that they have a unique contribution to make to our society. To distort our universities would be to deny ourselves this contribution. This is something we cannot afford to do, except in wartime when the alternative is even more dire. Hence, we need to look at the extent to which accepting DOD research money either has or necessarily must militarize our universities.

FIGURE 1 shows the total federal expenditures for academic research and development (R&D), in constant 1982 dollars, from 1967 to 1987.[4] Perhaps the most striking thing about the figure is that the National Institutes of Health (NIH), which supports medical research, towers head and shoulders above all other sources of funds. It would appear that academic R&D is more likely to be medicalized than militarized. Moreover, in most years the National Science Foundation (NSF) provided more funds than did the DOD, and in some years provided over twice as much as the DOD.[c] In addition, the funds from the other agencies shown always totaled more than the DOD funds.

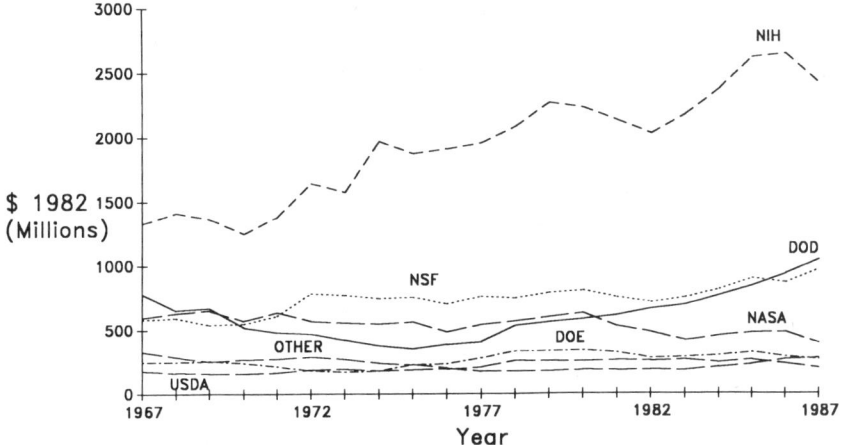

FIGURE 1. Federal funds for academic R&D, by agency.

FIGURE 2 shows the percentage of the total federal funding for academic R&D coming from each of the agencies from 1967 to 1987. The DOD never provided more than 20% of the total. The NIH always provided at least 30% of the total, and in some years nearly 40%. All other agencies combined provided between 40% and 50%. On the basis of these data, it would be hard to make the case that the DOD dominates the funding of academic R&D.

It might be objected that although DOD funding is only a small part of the total, there is a trend toward the DOD providing a greater share of the total. FIGURE 3 shows this objection is also invalid. From 1967 to 1987, total

---

[c] The drop in DOD funding during the 1970s resulted from an overreaction to the Mansfield Amendment, which limited DOD funding to research that was "relevant" to military needs. Many academic researchers, who evidently did not feel military contracts were a threat, complained bitterly when these funding cuts took place.

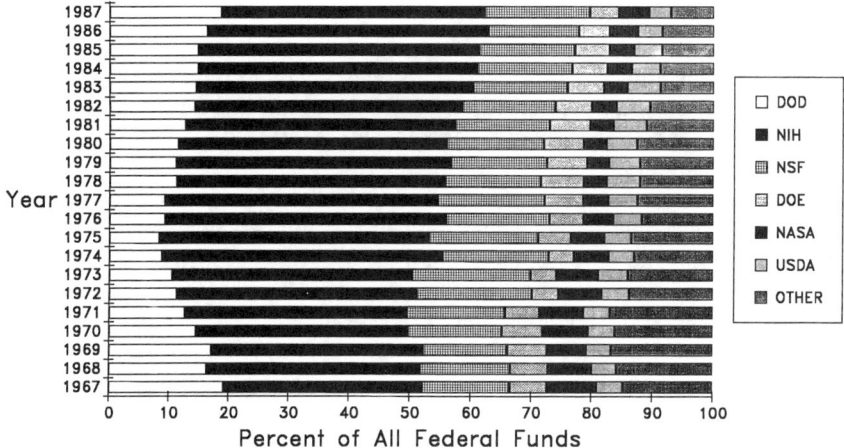

FIGURE 2. Agency percentages of federal funds for academic R&D.

federal funding for academic R&D grew, in constant 1982 dollars, at a compound annual rate of about 1.5%. The growth rates for both the NIH and the NSF exceeded 2%, while the growth rate for DOD funding was only about 1.4%. FIGURE 4 shows total growth in basic research funding from 1980 to 1988 (as defined by the American Association for the Advancement of Science).[5] Here we see that DOD funding grew less than did any other agency's. These figures show the DOD is clearly not on the way to domi-

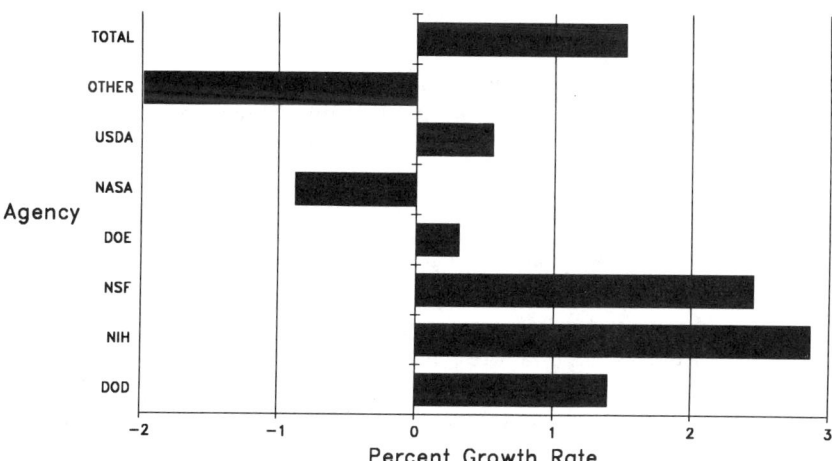

FIGURE 3. Real growth rates of agency funds for academic R&D, 1967–1987.

nating the funding of academic R&D. Militarization of universities, if it exists, does not come from overwhelming amounts of DOD funds.

Some critics argue that militarization comes, not so much from funding dominance, as simply from university involvement in the weapons development process, or as Rewak puts it, "the development, testing, and production of nuclear weaponry."

In addressing this argument, we need to make a distinction between *research* and *development*. The two tend to get lumped together, as in the phrase "research and development," but there are some important differences.

Research is a phenomenon-oriented activity. It is intended to gain an understanding of some phenomenon. The researcher may have any of a variety of motives for undertaking the research, ranging from pure curiosity to a desire to understand the phenomenon in order to solve some problem. Nev-

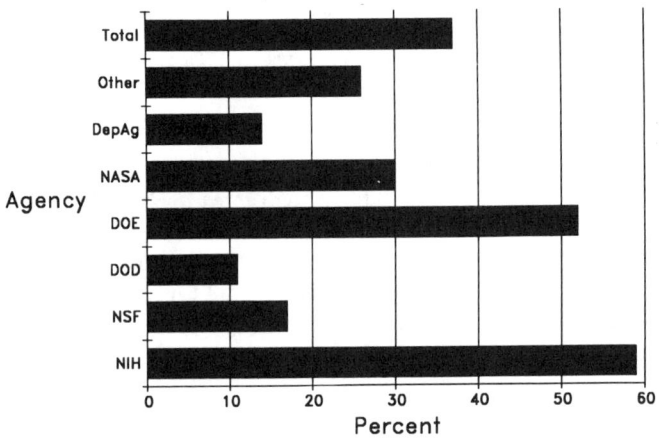

FIGURE 4. Real growth of basic research, 1980-1988.

ertheless, the focus is on the phenomenon. The researcher may abandon one approach to studying a phenomenon, and take up another that appears to be more fruitful. In his research, however, the focus of his activity is always on the phenomenon he has selected for study.

Development is a problem-oriented activity. The developer intends to solve some problem. He may make use of a variety of phenomena, abandoning some in favor of others, in the search for an optimal solution. The focus of his activity, however, is always on the problem to be solved, not on the specific phenomena he will utilize in solving the problem.

This distinction is important because virtually all DOD-sponsored "academic R&D" is research rather than development. The World War II work I referred to earlier, however, is an example of the latter. The MIT Radiation Laboratory developed several radars that went directly into production. Today there are a few universities that do involve themselves in development for the DOD and for other government agencies. Again, MIT provides an example. The C. Stark Draper Laboratory at MIT[d] has developed guidance systems for military missiles and for NASA spacecraft.

There are definite advantages to the conduct of development activities in universities. Engineering students, in particular, have the chance to learn from faculty members who are practicing engineering, not just teaching it. Some students are actually hired to work on these projects, and thereby gain experience they could not gain at universities refusing development contracts.

Likewise, there are definite disadvantages. Knowledge gained during a development project often only pertains to a specific technological problem. It may only indirectly apply to other problems, if it applies at all. The knowledge, even if not encumbered by security classification or proprietary restrictions, has little or no place in the technical literature or the classroom. It does not contribute to the university's purpose of generating and disseminating knowledge.

Some universities find the balance between advantages and disadvantages to weigh in favor of conducting development projects. Most, however, do not. There are very few institutions doing such work. Most universities receiving DOD funding conduct research, not development. Moreover, no universities develop nuclear weapons.[e] Hence, to describe what universities do with DOD funding as "weapons development" is misleading. To describe it as "nuclear weapons development" is downright mischievous.

To gain a deeper understanding of DOD-sponsored university research, we need yet another distinction, that is, one between *supporting* research and *purchasing* research.

Agencies such as the NSF and the NIH were established to *support* research in universities. These agencies do not themselves use or directly benefit from the research they pay for. Instead, the research is presumed to benefit the general public, and it is supported for that reason.

The DOD was established for a totally different reason: to defend the nation (the Preamble to the Constitution states that our government was established, in part, "to provide for the common defense"). In carrying out its mission, the DOD needs engineering and scientific knowledge. Therefore the DOD needs to *purchase* research that will enable it to carry out its mission.

As Kidd puts it, this purchase is "to ensure that an activity of great value to the nation as a whole is not neglected."[6] Kidd characterizes the purchasing of research as follows: "When [research] is purchased, the initiative

---

[d] This institution was known as the MIT Instrumentation Laboratory from the time of its founding by C. Stark Draper until his retirement.

[e] The Department of Energy's Lawrence Livermore Laboratory, which does design and develop nuclear weapons, is operated under contract by the University of California at Berkeley. Unlike Draper, which belongs to MIT, Livermore is really a part of the DOE, not a part of the University of California. Termination of the contract would mean that the DOE would seek another contractor, not that the University of California would redirect Livermore's activities. Moreover, many of the Livermore staff would probably go with Livermore to the new contractor.

lies with the purchaser. The purchasing agency first defines the problem it wishes to have solved and then seeks someone willing to solve it." By contrast, when the government supports research, "the initiative rests with the investigator. He indicates to the supporting agency what he wants to do. The supporting agency then assesses the investigator himself as well as what he proposes to do and decides whether support is warranted."

Weinberg makes a similar distinction.[7] "When a piece of research is done to further an end which society has identified as desirable, support for this type of scientific work should be considered as part of the bill for achieving the end, not as part of the 'science budget.' Only that scientific research which is pursued to further an end arising or lying within science itself should be included in our science budget."

The point of this distinction is that the motivations of the research buyer can be and often are different from the motivations of the researcher. Specifically, the DOD wants a particular piece of research for its own purposes, regardless of whether that research is likely to further progress in some branch of science. The researcher, on the other hand, will not undertake the research unless there is some chance of success, where success includes career-enhancing results. The DOD wants the research done because it intends to make use of the results. The researcher will do it only if it fits his own research agenda and that of his specialty.

What this means is that doing the research does not inevitably lead to militarization of the researcher or the university. There is a certain amount of distance between the researcher and the research buyer. Conducting the research involves a *quid pro quo*, a mutually beneficial exchange. As with any voluntary exchange, each party achieves his own goals by helping the other to achieve his. There is no more need for the researcher to adopt the mission-oriented goals of the sponsor than there is for the sponsor to adopt the scientific goals of the researcher. So long as both goals are legitimate, there is nothing unethical about the exchange.*f*

Having argued that accepting money from the DOD need not militarize either the researcher or his institution, I will now illustrate this point with some examples from my own institution. These examples show that research of high scientific quality can be done while enabling the DOD to solve some of its mission-oriented problems.

I have already mentioned the birdstrike work being done at my institution. It provides an excellent example of the kind of research I am discussing. Military aircraft, when flying at low altitudes and high speeds, risk collisions with birds. Although the risk for any single flight is small, the number of flights is so great that several birdstrike incidents can be expected each year,

---

f Note that I am *not* arguing that the distance between researcher and research buyer enables the researcher to "maintain his purity" while dealing with the DOD. We have already disposed of the idea that doing research benefiting the military is intrinsically unethical. The only issue remaining is whether such research weakens the university by militarizing it.

and military aircraft, which fly at high speeds, can be severely damaged. The Air Force typically loses a few aircraft each year to birdstrikes (in 1987, a B-1B crashed after colliding with a pelican). When the bird strikes the transparent canopy over the pilot, the pilot may actually be killed.[8]

Beginning in 1972, my institution won a series of contracts with the Air Force to do research on the birdstrike problem. It was a mutually beneficial exchange. The Air Force needed information about birdstrikes and how to cope with them, and we had the opportunity to conduct some very interesting research. Among the earliest results was a birdstrike probability risk assessment model. Developing this model required our researchers to extend standard statistical modeling techniques, and then to apply them to new problems. Another result was a finite element computer code for materially and geometrically nonlinear analysis (MAGNA), which is now a commercially available computer program. Our researchers experimentally characterized the loads resulting from bird impacts on aircraft canopies, and developed standardized test methods and procedures that have been accepted by the American Society for Testing Materials. Another outcome of the research was a simulated bird, which consists of a gel-filled plastic cylinder. This simulated bird, when fired against a test aircraft canopy, reproducibly duplicates the impact of a real bird, and allows us to conduct repeatable "birdstrikes" in the laboratory.

As a result of this work, our researchers have contributed to the knowledge of how materials behave during high-speed impact, and have received significant professional recognition on the basis of over 100 publications. The Air Force now has testing methods and design criteria for canopies that can withstand birdstrikes. Our research has resulted in fewer aircraft lost, and several lives saved. The reduction in aircraft losses alone more than paid for the research.

Another effort receiving a large share of its funding from the Air Force is our study of fatigue cracking of metals. Under normal operating stresses, metals undergo fatigue and develop cracks, which grow under repeated loads. The Air Force is concerned about two things. First, how small a crack is it possible to detect? Second, once a crack has been detected, how long will it take before the crack grows to the point were the structure fails? Answering these questions requires a great deal of innovative research. In conducting such research, we have developed new statistical techniques, and have advanced the state of knowledge of the fatigue behavior of metals. Our researchers have gained international recognition for their work. The Air Force is better able to predict the service life of its aircraft. Both have gained; neither has adopted the attitudes of the other.

The birdstrike work for the Air Force has an interesting postscript. We now have a contract with the Federal Aviation Administration to study bird

---

[8] To visualize the effect, imagine a chicken coming through your car windshield at 600 miles per hour.

collisions with civil aircraft. In civil aircraft, the primary problem is the ingestion of birds in jet engines. Worldwide, there are over 200 such incidents annually. The result is often damage to the engine, and in a few cases, loss of the aircraft and death of the crew and passengers. Our research on the bird ingestion problem will lead to design criteria for engines robust enough to survive bird ingestions, but not so heavy as to make the operation of aircraft too costly. We won this contract, at least in part, on the basis of our prior birdstrike work. An interesting sidelight of this research is that in analyzing the probability of damage from a bird ingestion, our researchers made use of a statistical technique we originally developed for the metal fatigue work. This is yet one more illustration that good basic research will have applications that cannot be foreseen at the outset.

Having said all this, do I then conclude that we need not be concerned about what kinds of DOD-sponsored research a university accepts? On the contrary, I assert there are certain ethical limits on the research a university should accept, and other limits that it may wish to impose on itself.

Just because the mission of the DOD is legitimate does not mean anything done in pursuit of that mission is legitimate. In wartime, it is necessary to interrogate prisoners to gain information. Torture, however, is generally considered unethical as a means of interrogation. The DOD has never requested research on scientific methods of torture. Should it ever do so, I believe the request should be rejected. Not, however, because such research is inappropriate for a university, but because such research is immoral for anyone. Nor is the problem restricted to the DOD. As a university with a Catholic tradition, my institution would not be likely to accept funds to conduct research on aborted fetuses, or on contraceptives, regardless of whether the money came from the DOD or the NIH. The research itself would be regarded as inherently immoral.

Moreover, there are other constraints a university may wish to accept in addition to the moral ones. There are certain kinds of work supported by the NIH that we would consider perfectly moral, but still inappropriate for us. We do not have a medical school. Thus we simply would not undertake research best conducted at a medical school. For us, it would be a blind alley. Likewise, we would not undertake a program of optical astronomy because of the light pollution inherent in our downtown location. In short, our location, our history, and our existing resources form a pattern of strengths and weaknesses. We will choose research that draws upon and adds to our strengths. We will avoid research that requires things we cannot have.

In summary, I argue that to ask about the ethics of DOD-sponsored research in the university is to get the question wrong. It implies either that there is something inherently unethical about the DOD, or that even if the DOD is not inherently unethical, there is something unethical about a university dirtying its hands with national defense. Attempting to draw the line between the university and the DOD is to put it in the wrong place. University faculty, staff, and students remain citizens and human beings. They do not lose their obligations as citizens. They do not escape the ethical obligations incumbent on all human beings. They have the same obligations as others not to engage in research that is intrinsically immoral. They do not incur ethical obligations not incumbent on others. They retain the same rights as others to undertake research that is not immoral, and that fits their interests and talents. Thus the proper place to draw the line is between immoral research and any researcher, regardless of whether he works in a university, in industry, or in a government laboratory.

Universities are attractive to research sponsors, especially the DOD, for two reasons: First, universities have no commercial interest in the outcome. Second, they do not involve themselves in sponsors' internal debates, as intramural laboratories sometimes do. Universities can be relatively objective in conducting sponsored research.

Conversely, universities have a great deal of freedom in choosing the kinds of research and sources of funds they will accept, and the conditions under which they will conduct the research. This freedom gives them as much protection as they want to have against being co-opted by their research sponsors.

Under these circumstances, it would be unfortunate for both parties if the DOD and our universities were prohibited from engaging in a mutually beneficial exchange. It would be especially unfortunate if this prohibition were motivated by unwarranted fears of the militarization of universities, or a false view of the ethics of national defense.

REFERENCES

1. REWAK, W. J. 1983. Universities and weapons research. Curr. Iss. Cathol. Higher Educ. 3(2): 27–28.
2. MARTINO, J. P. 1988. A Fighting Chance: The Moral Use of Nuclear Weapons. Ignatius Press. San Francisco, CA.
3. BURCHARD, J. E. 1948. Q.E.D.: M.I.T. in World War II. MIT Press. Cambridge, MA.
4. National Science Board. 1987. Science and Engineering Indicators–1987. U.S. Government Printing Office. Washington, DC.
5. ANON. 1989. The Reagan years: Mostly good for science funding. The AAAS Observer. January 6: 11.
6. KIDD, C. V. 1959. American Universities and Federal Research. Belknap Press. Cambridge, MA.
7. WEINBERG, A. M. 1967. Reflections on Big Science: 88. MIT Press. Cambridge, MA.

Ethics in the Pursuit of Ethics

DANNY COHEN

Information Sciences Institute
University of Southern California
Marina del Rey, California 90292

INTRODUCTION

I am grateful to the New York Academy of Sciences and Polytechnic University for inviting me to participate in this forum. I appreciate the opportunity to present my thoughts here.

I feel that having been myself a subject of investigation by the Office of Government Ethics (OGE), I have a valuable perspective on the subject. Incidentally, the investigation that was ordered by a congressional committee finally cleared me of any wrongdoing, but I do not want to discuss my case here, since it is not of general interest. However, the process itself is.

I apologize for using personal experience for some examples. Using personal experience is new to me. I never practice it in the papers and talks that I usually give, typically on computing and communications systems. Even if I am not fully qualified to discuss the subject, at least I have firsthand knowledge about it.

OUTLINE

This article touches first the question of what is ethics, and suggests that its main guardians are Congress and public opinion, through the media.

The article starts with an example of unethical conduct. This conduct, recently started in the government, is tolerated and considered acceptable because it is done in the name of ethics.

I argue that it is difficult to make objective ethical judgments, not because ethics is in the eye of the beholder, but because often what counts is not *what* is done, but *who* does it. For example, actions that are illegal and unethical when performed by some, are both legal and ethical when performed by others.

Congress's approach to ethics is not necessarily the same as that of public opinion and the media. An actual example is used to make this point.

Two different judgment systems are described: the court system and the congressional investigation system. These two systems are discussed and compared. Ethical pitfalls in the process are pointed out. These pitfalls, which are accepted when encountered in the pursuit of ethics, are amplified by the eagerness of the media to publish good stories.

This article concludes with two suggestions to reduce some of the unethical by-products that may accompany the pursuit of ethics.

WHAT IS ETHICS?

After consulting a few dictionaries, I concluded that ethics is compliance with a set of principles, accepted with or without being legislated. Under ethics, we expect compliance with a wider set of laws than that covered by legislation.

One can infer from this that whatever is illegal, meaning not within the laws of conduct, is also unethical. The opposite, however, is not necessarily so; that is, not all legal actions are also ethical.

It seems that often ethics is in the eye of the beholder.

The executive branch, through its Department of Justice, guards compliance with the legislated laws. But who ensures compliance with ethics? More particularly, who ensures compliance by the executive branch?

There are two guardians of ethics in matters related to the government, one is the OGE and the other is Congress, which through its many committees uses the process of congressional investigation. Because Congress is a guardian of ethics, it is also in a position to define what is ethical.

This article, in its discussion of the ethics of the process used to pursue ethics, raises several questions: Should the process used to pursue ethics also comply with the accepted ethics? Or is it above that? If it is not, who should guarantee that it is indeed conducted in an ethical manner?

THE SGE STORY

The following is an example of what I consider to be unethical conduct. This conduct, which recently began to be demonstrated by the government, is tolerated and considered acceptable because it is done in the name of ethics. All the details—the names, the subject area, and the government agency—have been changed to protect the guilty. The details are fictitious, but the essence of the story is true. It is a story about a situation that has occurred to several other people (but not to me!).

Dr. Smith is a world-renowned authority on metal coatings, particularly those used under water, and has directed a laboratory of a large corporation for over twenty years. He has recently succeeded in developing a special coating that could practically convert all the Navy's vessels into stealth platforms. After his invention is mentioned in the scientific literature, he is invited to attend a meeting of the Navy's Advanced Technology Panel, a request with which he is delighted to comply.

Three months after this meeting, Dr. Smith receives a letter from the Navy, thanking him for his participation. The letter also informs him, for the first time, that during the panel's meeting he should have been construed as a Special Government Employee (SGE), and therefore neither he, nor the laboratory he directs, nor any part of the corporation for which he works may seek any funding from the Navy for the next couple of years.

This is practically a kiss of death to Dr. Smith's laboratory.

The government's failure to mention anything about SGE status and its consequences before the meeting has no bearing on the case. In private, but never on the record, some government functionaries acknowledge that this was probably a slight oversight on the government's part, and in retrospect they can see why Dr. Smith is a little bitter about the whole experience.

This retroactive change of rules by the government is, in my opinion, both unfair and unethical. If it had been a retroactive change of anything else—an employee's salary, for example—this conduct would have been universally considered unacceptable. But because the change of rules with respect to SGE status was done in the name of ethics, it was considered acceptable by the government agency. No experienced government employee nowadays would dare try to correct this wrong, especially after consulting his department's legal counsel.

IS ETHICS IN THE EYE OF THE BEHOLDER?

Judging what is ethical is not simple. Laymen may err in considering particular situations.

Consider a hypothetical example: the Army, needing trucks, sends a young officer to the Oshkosh Truck Corporation to examine its products. Suppose, further, that during the visit his hosts serve breakfast, and that the officer does not pay for his share of the breakfast.

Is this legal? Ethical?

The correct answer is that this is neither legal nor ethical. In fact, this is a clear violation of both Department of Defense Directive 5500.7 (Standards of Conduct) and the law.

Consider another example: the Army, needing trucks, requests Congress (through the House Armed Services Committee) to approve the procurement of a certain number of trucks. Suppose that on the day when the committee meets to discuss this request, six members of that committee attend a breakfast given by the Oshkosh Truck Corporation, and not only do they not pay for that breakfast, but each of them is paid $2,000 for attending it, and suppose further that these members later on the same day pass a measure forcing the army to buy 500 more Oshkosh trucks than it needs.

Is this legal? Ethical?

The correct answer is that this is legal. But is it ethical? Some congressmen consider it ethical, and some do not. No laws or regulations are broken by this mere coincidental sequence of events. To the contrary, this is the norm for some.

Neither example is acceptable by public opinion, by the media, and by some congressmen. As an institution, however, Congress accepts the second example as ethical, and not the first.

Incidently, whereas the first example is hypothetical, the second is a factual description of an actual case. For example, *Time* reported on the case as follows: "Particularly questionable was a $2,000 payment on April 1, 1987, by the Oshkosh Truck Corp. to each of six House Armed Services Committee members just for coming to breakfast. A few hours later, an Armed Services subcommittee passed a measure to force the Army to buy 500 more Oshkosh trucks than it needs. Coincidence, says Oshkosh."[1] More details are available from the *New York Times*,[2] the *Washington Post*,[3] the *Los Angeles Times*,[4] and other sources. If this story is not true then I sincerely apologize to everyone involved.

In summary, the real question is not whether it is ethical for *any* of the government's officials to take money from contractors who directly depend on decisions made by these officials. The real question is *which* of these officials may take money, because it is illegal for some, but acceptable for those who are themselves in charge of guarding ethics.

Given the legality of such payments, it is no wonder that so many congressional committees involve themselves in procurement for the Department of Defense. According to a recent count, "29 committees and 55 subcommittees may oversee defense activities."[5] This year, according to the House Armed Services Committee,[6] the Senate/House conference on the National Defense Authorization Act for fiscal year 1989 "was the most crowded yet for a defense bill." The Senate Armed Services Committee was joined by 98 conferees from the House, including 36 members of the House Armed Services Committee, and 62 members of other House committees, some with unclear charters regarding defense acquisition, such as the House Education and Labor Committee, the House Energy and Commerce Committee, the House Merchant Marine Committee, and the House Post Office and Civil Service Committee.

THE TWO JUDGMENT SYSTEMS

There are two distinct judgment systems in the country: the court system and the congressional investigation system. The former convicts or clears the defendants; the latter, through public opinion and the media, "lynches" its victims. The courts are bound by rules of evidence, by the presumption of innocence until guilt is proved, and by similar technicalities. The congressional investigation system does not suffer from such handicaps. In courts, defendants are told what they are accused of, have access to the indictment, the right to defend themselves, and even the right to select their own witnesses. Not so in the other system.

The constitution defines the three branches of government and their prerogatives. The executive branch prosecutes and the judicial branch judges. Both branches participate in the court system of judgment. The legislative branch, or Congress, on the other hand, is free to use the other judgment system, the one based on public opinion and the media. In this system, Congress plays both the prosecutor and the judge.

In the congressional process, the rules of evidence have no role. Typically, there is the presumption of guilt until innocence is proved—to say the least.

We usually refer to McCarthyism when we wish to describe those instances in which *bad* congressmen (that is, congressmen with whom we disagree) use this process against *good* people (that is, people with whom we do agree). McCarthyism, in its full pejorative sense, is typified by men and women of zeal who do not scruple to advance their own (unworthy) causes by carrying out widely publicized and indiscriminate personal attacks. These men and women use means that are justified—according to their own narrow convictions—by the ends they seek.

But, how about when *good* congressmen use the very same process against *bad* people? Is it ethical then? Do good causes justify all means against people of different political persuasions than ours?

McCarthyism is a bad process, regardless of its political direction. It cannot be bad only when applied against those whose opinions we accept, and perfectly acceptable when applied against those with whom we do not politically agree. It is a bad process, in either direction. The Supreme Court Justice Louis D. Brandeis wrote in 1928: "The greatest dangers to liberty lurk in insidious encroachment by men of zeal, well-meaning, but without understanding."

Let me give an example of how the congressional investigation process does not adhere to the constitutional rights guaranteed under the Bill of Rights. I apologize again for using personal experience. The staff of the congressional committee that investigated me knew beforehand that I would be personally attacked in the hearing. Unlike a court of law, the congressional prosecutors/judges review in advance all the evidence that will be presented during the "trial." In addition, they tightly control what is presented, in a manner that may allow only one side of the story to be presented. Under this procedure, those accused are not given the opportunity to do anything equivalent to calling witnesses for the defense. Needless to say, I was not offered the opportunity to defend myself.

As a result of what was allegedly uncovered in the hearing, the committee asked the OGE to investigate me and two other people, and directed the OGE to address three particular charges. In a style typical to this process, the letter to the OGE (which was like an indictment) was made available to the press but not to those under investigation. Incidently, it took nearly five months for the OGE to clear us of these charges.

One of the limitations of the judicial process, as exercised by the courts, is that the indictment precedes the trial, which in turn precedes the verdict. The congressional/media process does not seem to suffer from this limitation. It is quite possible in this system for the press to publish in the morning what will occur in the afternoon hearing, who will testify to what effect, and most important, what conclusions the committee will reach, all based on information provided before the hearing. Time permitting, the defendants may, or may not, be informed of the charges, but most likely will never be given the opportunity to defend themselves. Hence, verdict first, trial later, with indictment that may never be posted. Somehow the congressional process reminds me of the immortal words of the Queen in *Alice in Wonderland*: "Sentence first—verdict afterwards."

This never occurs in courts. Judges do not provide the sentence and the verdict before the end of the trial, and they never use the technique of talking to the press "on condition of anonymity."

The Sixth Amendment guarantees the defendant's right "to be informed of the nature and cause of the accusation; to be confronted with the witnesses against him; to have compulsory process for obtaining witnesses in his favor." This holds only in courts of law, not in congressional investigations or the media. In a congressional investigation one may be found guilty without being told the nature of the accusation, and without being given the opportunity to defend oneself. In this process, the "prosecution" may bring forward any witness that might serve its cause. The "defense," however, is unable to bring forward any witnesses, and thus can hardly be said to exist. In addition, staffers routinely leak accusations against people, mostly of opposite political affiliations, while hiding behind the sacred condition of anonymity protected by the First Amendment, but ignoring the spirit of the Sixth Amendment, which is supposed to guarantee fundamental fairness in the process.

No wonder that the two senators from Maine, each of whom is from a different side of the aisle, classify "congressional investigation" as an oxymoron.[7] The Honorable William S. Cohen and the Honorable George J. Mitchell, the majority leader in the 101st Congress, have written that "in a court of law, the prosecutor and judge have different functions. A congressional committee is trapped in the unattractive dual role of prosecutor and judge."[8] This dual role, unattractive to honest people, is extremely dangerous in the wrong hands, as proved by the Gentleman from Wisconsin, Joseph R. McCarthy.

The rules that allow for congressional investigations probably should not be changed. They come with the territory defined by the separation of powers. We can only wish that this separation would be implemented more literally, or as meant by those who conceived of this separation in the first place. The system is basically good. Unfortunately, it is occasionally abused by some in a position to do so.

The abuses of the system one finds in congressional investigations are amplified by the eagerness of the media. A story is always newsworthy when someone is accused of illegal or unethical conduct. The fact that someone is not guilty is typically not newsworthy.

The same kind of abuse is not a monopoly of Congress. Being "under investigation" by the Department of Justice (DOJ) is also a precarious situation. It is reported by the media because it is news. If evidence is found to support the allegations, the case will be brought to court and reported in the media.

But what if there is no case? What if it becomes clear to the DOJ that the allegations are personally or politically motivated and lack any factual basis? The DOJ is not in the habit of issuing statements clearing people of wrongdoing. The best one can hope for is that the DOJ, in response to specific requests, will issue a statement such as "At this point in time there is insufficient evidence to bring the case to court." Hardly a closing of the investigation. Hardly a clearing of one's name. Hence, if there is no case, the investigation is typically an indefinite process, with no official end, ever.

Because an investigation may be kept open indefinitely, and because at least 99 out of 100 lawyers would advise against making public statements, a person under investigation may never be able to present his side of the story by talking to the media. To make things worse, disappointed reporters routinely make a "no comment on advice of my counsel" to read like an admission of guilt.

This, too, should not be changed. It comes with the territory of the freedom of the press. We can only wish that information would be conveyed to the public in a more responsible and fair manner. The system is basically good. Unfortunately, it is occasionally abused by some in a position to do so.

About 20% of those who are accused and brought to the various U.S. District Courts in criminal cases are acquitted, and have their reputations cleared (in the year ending June 30, 1986, out of 50,040 defendants in criminal cases, 9,300 were either acquitted, by jury or court, or had their cases dismissed[9]). What percentage of those accused through the process of congressional investigation get the equivalent of acquittals and having their reputations cleared? Given that the budget of the DOJ is so much larger than that of most congressional investigations, and that the DOJ, unlike Congress, employs professional investigators (including agents of the FBI) and professional prosecutors, we should assume that the percentage of successful prosecutions in the judicial process is higher than that in the congressional process. I cannot recall, however, a congressional committee issuing a statement clearing anyone mistakenly accused of any wrongdoing. Because congressional investigations combine the roles of prosecutor and judge, I dare suggest that they have a moral and an ethical obligation to issue statements that would be equivalent to the acquittals granted by the courts.

SUGGESTIONS FOR REFORM

I have two suggestions, motivated both by my own experience and by my reactions to similar experiences others have had.[a] The first, which is for the government, addresses the problem that investigations may last forever, without a cause. The second, which is for the media, is about parity in reporting on investigations.

---

[a] For a dramatic treatment of trial by media, see Kurt Ludke's *Absence of Malice*. Kerry Stewart has adapted the screenplay into a book, with the same title, published by Ballantine in 1982.

Time Limits for Investigations

I suggest that any investigation, by any branch of government (legislative or executive), should have a defined time limit. Beyond this limit, an investigation may be extended by a judge if cause is shown. Otherwise, either a case must be brought to court or a statement issued clearing the investigated person. If new evidence is discovered, an investigation may be reopened at any time, with a new time limit. The purpose of this suggestion is to make sure that no one is kept in the precarious state of being under investigation forever. This is most important for investigations that have been reported in the media, either through open statements or through leaks.

Parity in Reporting

I do not believe that the media should be told how to report anything, including investigations, whether conducted by the DOJ or by Congress.

Any news organization that reports any accusations, however, has a moral and an ethical obligation to report also the clearing of these accusations, in the same conspicuous manner as the original report about the accusations was made.[b] The same should also hold for indictments and acquittals (including case dismissals, for any reason), and for the opening and closing of investigations.

Experience proves that this moral and ethical obligation is not generally observed, probably because stories about acquittals sell fewer newspapers than stories about accusations. Therefore, I suggest that this should be a legal obligation, whether the media find acquittals newsworthy or not, and whether the media would expect acquittals to be of interest to their readers or not. This is not any infringement on the freedom of the press, or a form of censorship. It is just a call for fair play, similar to the equal time requirement.

CONCLUSION

The process of pursuing ethics should not victimize anyone, including those holding opinions and beliefs that are different from our own. Victimizing anyone would not be ethical, even when done in the name of ethics.

When James F. Beggs was the acting director of the National Aeronautics and Space Administration (NASA), he was accused of defense contract fraud that had occurred while he was working, earlier, for General Dynamics. The

---

[b] Any medium (for example, a newspaper, a magazine, or a television station) that does not report an accusation is under no obligation — according to this suggestion — to report any subsequent acquittal or dismissal.

charges, which were made not long after the Challenger accident, received extensive coverage by the media. The media all but ignored his case, however, when he was subsequently cleared of these charges. Although the Attorney General later took the extremely rare action of issuing a formal apology, and although this apology did receive press coverage,[10,11] Mr. Beggs can hardly be said to have been treated fairly. To paraphrase Mrs. Beggs: When you are indicted it is on the front page, but when you are cleared it is in the obituaries, if at all.

This happens frequently. It is legal, but not ethical, and should be changed.

REFERENCES

1. CARLSON, M. 1988. The foul stench of money. Time. July 4: 21.
2. HERSHEY, R. D., JR. 1988. Members' conduct may be in line but out of step. New York Times. June 21.
3. KNIGHT, J. 1988. Give us that old-fashioned "honest graft." Washington Post. June 21.
4. CARLTON, J. 1988. $2,000 to each from supplier: Badham [R-CA], 5 colleagues took money before vote. Los Angeles Times. June 11.
5. ADAMS, J. A. 1988. How the Pentagon buys. IEEE Spectrum. November: 29.
6. House Armed Services Committee. 1988. News release. July 7.
7. ENGELBERG, S. 1988. Undone by Oliver North. New York Times. September 25.
8. COHEN, W. S. & G. J. MITCHELL. 1988. Men of Zeal: A Candid Inside Story of the Iran-Contra Hearings: 161. Viking. New York, NY.
9. U.S. Department of Justice. 1987. Sourcebook of Criminal Statistics, 1987: 442. Bureau of Justice Statistics, U.S. Department of Justice. Washington, DC.
10. Editorial. 1988. An inadequate apology. Houston Post. July 9.
11. Associated Press. 1988. Apology for ex-NASA chief. New York Times. July 7.

Military Research, Secrecy, and Ethics[a]

VIVIAN WEIL

Center for the Study of Ethics in the Professions
Illinois Institute of Technology
Chicago, Illinois 60616

INTRODUCTION

Up to now, scientists and engineers have not shown much interest in the right of free expression as it applies to the dissemination of their research. One specialist makes the point this way: "Scientists are not a First Amendment consumer group." Journalists, of course, are the prime example of a First Amendment consumer group.

Scientists generally invoke two principles to support their claims to open dissemination of scientific and technical information: 1) the norm of openness for the enterprise of science and 2) academic freedom for university research. The norm of openness in science rests on assumptions about the nature of scientific advance and discovery. Independent critical scrutiny and replication of findings are thought to be essential. Restrictions on the flow of information interfere with processes by which scientists' results are built upon or challenged by others.

In the university context, the norm of academic freedom supports open dissemination of scientific and technical information. Restraints on communication hamper the university in performing its two main functions, research and teaching. Moreover, free exchange within an academic community not only provides a model for the wider society but has intrinsic worth; it is valuable in its own right.

These rationales are very familiar, and scientists generally view them as adequate for protecting their interests in open dissemination. One might go so far as to say that scientists have been averse to characterizing the values at stake as First Amendment rights, as issues of free expression. When they encounter threats to openness from government restrictions, scientists appar-

---

[a] This essay derives from a current research and writing project directed by the author and supported by a grant from the John D. and Catherine T. MacArthur Foundation. The project, which is entitled "National Security, the First Amendment, and Scientific and Technical Information," brings together research by six other investigators. The discussion in the above essay draws primarily from an unpublished manuscript by one of the investigators, Benjamin DuVal, Jr.; from personal discussion with another investigator, Robert Ladenson; and from the latter's unpublished manuscript. The manuscripts of all the investigators will be published when the project is completed.

ently prefer to negotiate arrangements with government officials than to invoke First Amendment rights.

In the last year and a half, controversy has abated, with the previous administration having retreated from some restrictive initiatives and scientists having become acquiescent. It is nevertheless important to explore the implications of the right to free expression for communication of scientific and technical information. Regulatory authorization now "in dormancy" can be activated when the political climate shifts. If scientists have been shy about invoking First Amendment rights because of their dependence on government funding, there is all the more reason to investigate what restraints the First Amendment places on government control of the flow of technical information. Although thinking about First Amendment issues is alien to scientists, it may yield intellectual as well as practical rewards. The discussion that follows focuses on the value of free expression, but the need to accommodate other values, such as our national security interests, figures importantly in the analysis.

FREE EXPRESSION: THE UNDERLYING RATIONALE

My plan here is to scrutinize ways of categorizing scientific research and researchers as a basis for determining the implications of First Amendment protection. The aim is to get a clearer picture of the scope of legitimate government restraints on scientific communication. Studying the rationale for First Amendment protection of expression helps us understand the implications of this constitutional right for the transmission of information, specifically scientific and technical information.

The classic case for free expression appears in the second chapter of John Stuart Mill's work *On Liberty*. Contemporary formulations of the rationale emphasize two chief considerations. The first is the necessity of an open flow of speech to foster the growth of individual and collective rationality and the discovery of truth. The second is the indispensability of free speech if citizens are to participate in government and hold government officials accountable for their actions. Scholars do not agree on whether these two strands of justification can be shown ultimately to amount to or rest on a single principle.

A legal scholar, Benjamin DuVal, Jr., holds that these strands are distinct and contends that they do not necessarily have the same implications for the scope of protection of dissemination of information from government restrictions. He believes that the democratic values of citizen participation and public accountability of government officials require greater protection of dissemination of certain information than does the value of fostering rationality.

From historical experience, we conclude that there is a close link between the development of democratic political processes and the demand for information about government affairs. Citizens need information about the actions of government through its officials to gain a more equal footing on which to judge and criticize government policies. In addition, communication may be thought of as part of a checking mechanism against misconduct on the part of government officials. Restrictions on the flow of information about government affairs and the conduct of government officials present risks that differ importantly from the risks of restrictions on other kinds of information. Secrecy about government affairs can cloak incompetence and corruption and discourage a lack of care in decision making. By reducing the number who are aware of what is being done and discouraging disclosure by those "in the know," secrecy reduces scrutiny from inside and outside. Officials protected by secrecy may undertake enterprises that they would otherwise avoid out of embarrassment or fear that the public would not tolerate them.

Secrecy has a tendency to spread. This is in part because control of information confers the power to shape public attitudes toward government policies. Secrecy can also spread because it often takes less time and effort to stamp "classified" on an item that may conceivably warrant classification than to subject it to careful review to see whether classification is indeed called for. There are various estimates of the extent of overclassification.[1] It is difficult, if not impossible, for the outside observer to assess how much overclassification there is.

These considerations come into play when the information in question concerns government actions or reflects directly on the wisdom of government policies. This is not to say that other information is irrelevant to political decision making. DuVal's point is that in the case of information that has no direct relation to the assessment of government policies, the risks are not as great that government officials will abuse their positions.

DuVal asks us to compare the costs of restricting research into genetic engineering with those of restricting the dissemination of results from weapons tests. Restricting genetic engineering research might reduce the fund of knowledge that could be called upon in decision making, and it might make certain options unavailable to society in the future. Thus, there may well be political consequences. Denying this knowledge, however, does not reduce the ability of citizens to take part in political decision making or their ability to hold public officials accountable for their conduct. Moreover, the reasons public officials would have for restricting such research do not derive from their interests as public officials. This means that the risks that they might abuse their position are probably not as great in this connection. In the case of weapons tests, however, officials have reasons to keep embarrassing failures secret or to release only favorable results regarding their own favorite programs. Their views about what are correct policy choices are not a legitimate basis for restricting information. In a democratic society, the notion that releasing certain information about weapons tests, for example, might lead some citizens to choose wrongly (from the officials' perspective) does not count as an appropriate reason for restriction. Of course, officials usually offer reasons that do not directly counter democratic values. They appeal to the need to keep the information from potential adversaries so that the latter are not informed of the deficiencies or advantages of existing or proposed weapons systems.

LEGITIMATE RESTRICTIONS ON INFORMATION

Continuing this line of reasoning, which is based on the importance of a free flow of information to preserve democratic values, the next question is what are legitimate grounds for limiting the flow of information about government actions. To help answer this question, Sissela Bok has proposed a distinction between two levels of government secrecy.[2] "First order secrecy" concerns the actions of government officials, the reasons for their actions, and the information needed for assessing their actions. "Second order secrecy" concerns the fact that certain information or a category of information is being kept secret, the reasons for secrecy, and the information necessary for assessing the need for secrecy. What cannot be justified is second order secrecy. This means that it is illegitimate to restrict information that tends to show that secrecy is unnecessary or undesirable. Such restriction disables critics and thereby obstructs accountability.

It is not clear, however, whether the distinction between these two levels of secrecy can be maintained in practice. DuVal points out, for example, that the Church Committee hearings showed that the most effective argument against covert operations may well be the disclosure of these operations. Similarly, the distinction between information about government action and information that bears on the justifiability of government action breaks down. One can regard the same information about the costs and capabilities of a weapons system alternatively as part of a description of the action of acquiring the system or as information relevant to determining whether the acquisition is advisable.

Yet there must be some legitimate restrictions on the dissemination of information about government affairs. Indeed, some restrictions do not intrude upon democratic values. DuVal suggests two categories of information that can be safely suppressed. One category consists of information at the level of details that have slight relevance to political debate. Another category consists of issues that, by common agreement and practice, are not appropriately settled by political debate. Restrictions on these categories of information can therefore be justified.

The category of details would include, for example, some technical data on weaponry. Presumably, the public can adequately assess the costs and capabilities of a weapons system without access to all the details of the design. An example of an issue not to be settled by public debate is the planning of a military operation before the fact. After the fact, information must be disclosed, for it is critical to evaluating how the operation was carried out.

The question of whether to undertake a military operation at all, however, may not be a matter excluded from settlement by public debate. Public discussion of this question is not so obviously self-defeating in all cases.

There remains the dilemma of how to deal with scientific or technical information that is important for democratic decision making and at the same time valuable to potential adversaries. To respond, DuVal proposes some categories of information that government may not safely be allowed to suppress. As we have seen, second order secrecy is not justifiable: information that a category of information is being kept secret should not be denied. Citizens cannot assess the wisdom of a weapons system if they do not know the weapons system exists. Although a government should not keep secret the existence of an antisatellite program, for example, it may justifiably restrict information about the results of tests or details of design of particular systems.

There are, DuVal suggests, other "safe harbors" of information that should always be disclosed. They include information revealing a violation of law and, more controversially, general information about the nature, capacity, and costs of weapons systems. For the rest, what may serve best, in the view of a number of scholars, is an administrative government procedure for reviewing information on a case-by-case basis, under general standards. The standards should be so devised as to ensure that citizens' rights (to information needed for participating in policy decisions and holding officials to account for their actions) will be taken into account as well as dangers to national security. Another scholar, Robert Ladenson, suggests that the appropriate general standard for case-by-case review may well be the "clear and present danger" test.

LEGITIMATE RESTRICTIONS ON GOVERNMENT EMPLOYEES

An important legal case, *Snepp* v. *United States*, upheld the authority of the government to impose prepublication review requirements on employees of the Central Intelligence Agency who have access to highly sensitive material.[3] This decision was based on the assumption that the government can put restrictions on employees that it cannot impose on the general public. How far government can go in imposing such a requirement on government employees, contractors, or grantees outside the intelligence community has not been tested. The extent of the government's prerogative in this regard should be of keen concern to scientists and engineers, among others.

That government has greater latitude to impose restrictions on its own employees is widely conceded. The proposition that government has to make a far stronger case to impose restraints on individuals outside government in regard to information neither generated nor controlled by government receives strong endorsement. Indeed, all influential spokespersons from the academic and scientific communities expressed this view when they responded to restrictive initiatives of the Reagan Administration. (These initiatives included, for example, government interference with communication between American and foreign scientists in open university laboratories.) It is difficult, however, to determine the basis for making this important distinction between restraints that apply only to individuals connected with the government (and concerning government-generated or -controlled information) and restraints that apply to anyone at all (and concerning information generated and controlled outside government).

How should we make sense of it? Robert Ladenson suggests that we can understand this contrast on the basis of the considerations scientists traditionally appeal to, the advance of science and an open environment for research and teaching in the universities. He suggests that the work of physicists engaged in weapons research at Los Alamos or Lawrence Livermore Laboratory, for example, may not matter significantly to the advance of science or the academic environment.

Questions remain, however. Are we ready to say that the quantity of work carried out in such restrictive settings matters little to the advance of science? Should we not recognize that boundaries often become blurred? (Scientists at Lawrence Livermore Laboratory are not government employees; they receive their salaries from the University of California.) Do we not benefit from communication among scientists, no matter who sponsors their work?

Even more challenging is the rationale for the contrast from the standpoint of Mill's defense of free expression. Perhaps restricting the dissemination of research generated or controlled by government is seen as merely declining to release to others information that one possesses, whereas restricting the dissemination of research generated or controlled outside government, and obliging individuals to keep information secret, is regarded as preventing a person from speaking. As Ladenson observes, at the level of interpersonal conduct, this distinction has moral significance. If individual A prevents individual B from speaking, all other things being equal, A wrongs B. On the other hand, if A simply declines to provide B desired information, then again, all other things being equal, A does B no wrong, or at least less wrong.

Ladenson notes, however, that when the parties to this relationship are government and individuals subject to its authority, the distinction between not releasing information and preventing someone from speaking becomes blurred. This is because governments are frequently the only source of information about important matters related to public affairs. Think of how the early stages of the invasion of Grenada were concealed. In such circumstances, when government chooses not to disclose certain information, there is an implied judgment about what the public needs to know. This seems essentially similar to judgments associated with government restraints on the flow of information between individuals outside government. In both sorts of circumstances, according to Ladenson, the judgments are not much different from those a censor would make.

Yet, we must concede that government does have a claim on its own employees (and on information it generates or controls) that it does not have on the general public (and on information outside its purview). Some would say it *owns* its information, and can protect this information as companies protect trade secrets. Whether this is an appropriate comparison or not, the government's claim is not unlimited. To define the extent of this claim, we are again led to look for a structuring process, for an adequate institutional mechanism. Such a mechanism is necessary if we are to avoid, for example, sweeping prepublication review agreements. The prospect of government decision makers individually balancing, to the best of their ability, the interests of national security against the needs for open flow of information drives us to search for a procedure that would ensure that these competing values will be considered in good faith.

The advantages of devising appropriate institutional procedures, in spite of the costs, become evident from surveying the costs of secrecy in the past. In a recent review of McGeorge Bundy's book, *Danger and Survival: Choices about the Bomb in the First Fifty Years*, Professor Stanley Hoffman of Harvard University emphasized Bundy's case against the government's penchant for secrecy. Although secrecy was justified in World War II, the first bomb project set a dangerous pattern, in Bundy's view. "The commitment to secrecy delayed and restricted to an exceedingly small number of people the debate on dropping the bomb. . . . The obsession with secrecy prevented a public discussion of the move to the H bomb, . . . [and] it held Eisenhower back from explaining why the alarmist warnings of a 'missile gap' favoring Russia, after the launching of Sputnik in 1957, were wrong." According to Hoffman, Bundy supplies a number of other examples. He says that Bundy now finds the reasons for secrecy in these cases less convincing and regards the secrecy as a cause of mistakes. With hindsight, Bundy sees such secrecy as deception practiced on colleagues, the American public, and United States allies. He is equally stern in his judgment of the effects of secrecy on the Soviet side.[4]

This seems an apt time to pursue further the effort of indicating the limits that our commitment to free expression puts on government interference with the flow of scientific and technical information. It may also be a good moment to begin devising administrative procedures that would weigh both the harm from disclosure and the importance of the information to the political process. By such means, we may save ourselves from excessive secrecy and the damage it does to the welfare of our society.

REFERENCES

1. For recent contrasting estimates, see Information Security Oversight Office. 1984. Annual Report to the President (FY 1983). See also Hess, S. 1984. The Government/Press Connection. Brookings Institution. Washington, DC.
2. Bok, S. 1983. Secrets: 112–115, 201–205. Vintage Books. New York, NY.
3. Snepp v. United States, 444 U.S. 507 (1980).
4. Hoffman, S. 1989. Danger and survival: Choices about the bomb in the first fifty years. N.Y. Rev. Books **36**(1): 29.

Scientific Freedom and Scientific Responsibility

HAROLD C. RELYEA[a]

Congressional Research Service
Library of Congress
Washington, DC 20540

INTRODUCTION

From a functional perspective, scientific freedom prescribes maximum individual autonomy for each practitioner in selecting research subjects, communicating with colleagues, and validating or invalidating scientific findings. Extending knowledge and improving its soundness is possible in science, and to some degree in technology, to the extent that individual autonomy is unrestrained.[b] Science and technology, however, are not pursued in a social and political vacuum.

Investigators of the mysteries of nature, from at least the dawn of Christianity to the eighteenth century, were well aware that their freedom to conduct research and report resulting discoveries was subject to limitation by the ecclesiastical and civil authorities of the day, as well as by the baser instincts of society. Fear of reprisal by the church, the state, or the mob was adequate incentive for initially cloaking scientific activities in secrecy. As historian Carl Becker has commented, however, much had changed by the dawn of the eighteenth century—"the moment in history when men experienced the first flush and freshness of the idea that man is master of his own fate."[1] Science and technology contributed significantly to this conviction. The workings of nature were but expressions of the will of God; to fathom the phenomena of nature was to reveal the beneficent intentions of the Almighty. The discovery of these truths through human reason and the accumulation of verifiable knowledge obviated the need to continue to rely upon official and dogmatic pronouncements of the church and the state.[2]

Subsequently, however, as science and technology advanced, announcements about Holy revelations deriving from the analysis of nature were forsaken. In fact, as research results provided humankind with power over things, and even provided some human beings with power over other human

---

[a] The views expressed here are solely those of the author and are not attributable to any other source.

[b] Suffice it to say here that *science* is understood as encompassing those endeavors in which knowledge about the physical world is sought and codified, whereas *technology* is understood as encompassing those endeavors in which knowledge is applied so as to alter and control the material conditions of the world.

beings, it became increasingly apparent that discoveries could be used indifferently for any purpose, whether good or bad. Ironically, while science and technology flourished in an atmosphere of freedom, their fruits could very well be utilized to restrict or destroy that quality.

Indeed, scientific and technological results gave no allegiance to any particular form of government. Consequently, constraints have been set upon scientific freedom through law as well as official policies and practices by various governments, including our own.[3] In the United States, national security controls exist apart from security classification restrictions, which may be applied to research and development efforts sponsored or underwritten by the federal defense community, and are applicable to scientific and technological knowledge in which the federal government has no proprietary interest or right of ownership.[4]

LIMITING SCIENTIFIC FREEDOM

Among these controls is the Invention Secrecy Act, which provides that the Patent Commissioner, "whenever the publication or disclosure of an invention by the granting of a patent, in which the government does not have a proprietary interest, might, in the opinion of the Commissioner, be detrimental to the national security," shall make the application available to certain specified defense agencies for review. In the event that one of these reviewing entities determines that "the publication or disclosure of the invention by the granting of a patent therefor would be detrimental to the national security, . . . the Commissioner shall order that the invention be kept secret and shall withhold the grant of a patent" for not more than one year, subject to a possible renewal.[5] These secrecy restrictions may be appealed to the Secretary of Commerce,[6] and a claim for compensation for the damage caused by such a secrecy order may be made through the proper federal court.[7]

An inventor subject to such an order who willfully publishes or discloses the information it covers not only forfeits his or her patent right, but also can be fined $10,000 or imprisoned for two years, or both.[8] The proprietary information in question is not usable. Compensation is difficult to obtain, and usually no details are provided regarding the national security detriment prompting the government's action.[9]

A few years ago, when a patent was sought on a cipher device perfected by Dr. George Davida, an associate professor of electrical engineering and computer science at the University of Wisconsin–Milwaukee, a secrecy order, sponsored by the National Security Agency (NSA), resulted in April 1978. Considerable adverse publicity was engendered, however, and the order was lifted in June.[10]

At about the same time, three inventors seeking a patent for a voice scrambler that would let radio and telephone users converse without being overheard by others were also met with a secrecy order instigated by the NSA.

Once again adverse publicity ensued. The secrecy restriction was suddenly removed several months later, after one of the inventors charged that the order "appears part of a general plan by the NSA to limit the privacy of the American people." "They've been bugging people's telephones for years," he said, "and now someone comes along with a device that makes this a little harder to do and they oppose this under the guise of national security."[11]

The director of the NSA subsequently told congressional overseers that these situations exemplified "not a faulty law but inadequate Government attention to its application."[12] Nonetheless, other secrecy order recipients have not fared quite so well, perhaps because they felt incapable of mounting successful challenges to similar national security restrictions. Moreover, secrecy orders not only suppress particular inventions and frustrate the inventors responsible for these inventions, but can also discourage scientists and technologists from pursuing work in some areas because of the likelihood of secrecy orders being imposed.

Relying upon the national security provisions of the Export Administration Act, government efforts to regulate the communication of certain scientific information through professional publications and conferences began around February 1980.[13] At that time, the Department of Commerce informed the organizers of a small international meeting, which was to be held under the auspices of the American Vacuum Society, and which was to discuss magnetic bubble memory devices, that the presence of certain foreign nationals at the gathering subjected the scheduled proceedings to compliance with export requirements. In order to avoid possible fines or imprisonment for failure to comply with export regulations, the organizers of the meeting rescinded the invitations that had been sent to scientists in Hungary, Poland, and the Soviet Union. Scientists from the People's Republic of China, who were already en route, were permitted to attend after signing an agreement not to "re-export" what they learned to any nationals from eighteen named countries.[14]

In April 1984, the Committee on Scientific Freedom and Responsibility of the American Association for the Advancement of Science released a compilation documenting this incident and a dozen similar incidents that involved professional societies.[15] In one case—the Twenty-Sixth Annual International Technical Symposium of the Society of Photooptical Instrumentation Engineers—more than 100 technical papers scheduled for presentation were withdrawn at government insistence.[16-18]

In May 1984, individuals attending the Twenty-Fifth Structures, Structural Dynamics, and Materials Conference found that two of its sessions were closed to foreigners. The conference program carried a notice citing the International Traffic in Arms Regulations as the basis for the restriction, and advising those seeking admission "you must have with you a document signed by your company security officer (e.g.: company badge which specifically states U.S. citizenship), birth certificate, naturalization papers, passport, or voter registration card in order to certify that you are a U.S. citizen." "If you fail to carry this citizenship material with you," it continued, "you will be denied admission to these sessions."

This self-censorship occurred even though one of the conference sponsors, the American Institute of Aeronautics and Astronautics, had a policy of holding open, public meetings if at all possible. Reporting on the incident, *Defense Week*, a privately published newsletter, commented as follows: "Never before have organizers volunteered to close a meeting unless it was jointly sponsored by the federal government."[19]

It soon appeared, however, that other scientific professional societies, perhaps anticipating government intervention under the export laws, were willing to close some of their conference proceedings to all but U.S. citizens. Furthermore, on the campuses, both the University of Maryland at College Park and the University of California at Los Angeles demonstrated they were willing to restrict a short course on metal matrix composites, taught in cooperation with the Metal Matrix Composites Information Analysis Center, to only U.S. citizens because some composites are export controlled.[20]

THE NATIONAL SECURITY PROBLEM

The central regulatory value underlying these restrictions on scientific communication is *national security*, a nebulous term of shifting meaning and changing understanding in federal law and public policy. That national security is not always susceptible to definition, interpretation, or understanding by an individual or a collectivity may be both its strength and its weakness as a policy concept. On the one hand, its imprecise and general nature can provide considerable flexibility for government action; on the other hand, its vagueness may render the term senseless or subject to accidental or willful misinterpretation, resulting in ventures that are detrimental to other rights and privileges of the polity.[c]

On its face, national security concerns the security of the nation. There is a considerable variety of opinions, however, as to what terms of reference or perspectives the concept embraces. The Department of Defense reportedly has described it as "the condition provided by: *a.* a military or defense advantage over any foreign nation or group of nations, or *b.* a favorable for-

---

[c] Regarding the latter consideration, Egil Krogh, chief of the White House plumbers' unit, expressed the following comment at the time of his sentencing for perjuring himself concerning the burglary of the office of Daniel Ellsberg's psychiatrist:

> I see now . . . the effect that the term "national security" had on my judgment. The very words served to block critical analysis. It seemed at least presumptuous if not unpatriotic to inquire into just what the significance of national security was. . . . The discrediting of Dr. Ellsberg, which today strikes me as repulsive and an inconceivable national security goal, at the time would have appeared a means to diminish any influence he might have had in mobilizing opposition to the course of ending the Vietnam War that had been set by the President. Freedom of the President to pursue his planned course was the ultimate national security objective.[21]

eign relations position, or *c*. a defense posture capable of successfully resisting hostile or destructive action from within or without, overt or covert."[22] By contrast, a recent official Japanese report discussing "comprehensive national security" eschews military terms for predominantly economic considerations.[23]

There have been other explanations of national security, reflecting perspectives on the concept of varying breadth and diversity. For example, one political scientist, Gabriel Almond, viewed national security policy as "that part of a nation's foreign policy which is concerned with the allocation of resources for the production, deployment, and employment of what we might call the coercive facilities which a nation uses in pursuing its interests."[24] Another, Samuel Huntington, has indicated it is meant "to enhance the safety of the nation's social, economic, and political institutions against threats arising from other independent states."[25] General Maxwell Taylor, who, by his own admission "spent most of his adult life in activities related to national security," explained the concept in terms of "national valuables," which include "current assets and national interests, as well as the sources of strength upon which our future as a nation depends." According to General Taylor, some of our assets are "tangible and earthy," and others are "spiritual or intellectual," and "range widely from political assets such as the Bill of Rights, our political institutions and international friendships, to many economic assets which radiate worldwide from a highly productive domestic economy supported by rich natural resources." In his view, "the urgent need to protect valuables such as these . . . legitimizes and makes essential the role of national security."[26]

National security policy analysts Harold J. Clem and Stanley L. Falk view a state's national security as "something to be defined within the context of its external environment and the international order." They contend that "in its search for security the state endeavors to relate with its external environment in a way conducive to the preservation of those political, social, and economic institutions which it values." They also contend, however, that security "is a measure both of the absence of actual threats to acquired values, and of the absence of fear that such values are threatened." Hence, "security is to be seen partly as a state of mind of a people and of its leaders." Furthermore, according to Clem and Falk, this state of mind need not be the same for both groups. Consequently, among the most influential factors, "national character, tradition and prejudices play most important roles in shaping a nation's security outlook."[27]

More recently, former White House national security adviser Richard V. Allen offered a characterization of national security, saying "it must include virtually every facet of international activity, including (but not limited to) foreign affairs, defense, intelligence, research and development policy, outer space, international economic and trade policy, monetary policy, and reaching deeply even into the domains of the Departments of Commerce and

Agriculture." For this practitioner, national security "must reflect the presidential perspective, of which diplomacy is but a single component."[28]

Political scientist Richard Ullman advanced the contemporaneous view that "a threat to national security is an action or sequence of events that (1) threatens drastically and over a relatively brief span of time to degrade the quality of life for the inhabitants of a state, or (2) threatens significantly to narrow the range of policy choices available to the government of a state or to private, nongovernmental entities (persons, groups, corporations) within the state."[29]

These viewpoints reflect very little consensus about the terms of reference of the national security concept, except that it includes a military dimension. In terms of the dispute over recent government efforts to exert greater control over scientific communication, opponents contend that too much scientific and technological knowledge is seen to have military value and, therefore, is deemed to warrant national security restriction. This view was clearly reflected in the 1982 report of the National Academy of Sciences' Panel on Scientific Communication and National Security. After urging federal officials to "concentrate on the most *feasible forms of control*" and "eschew regulations that impose compliance burdens without significantly affecting leakage," the Panel indicated that the government "should concentrate its resources more systematically on those *technologies that are of greatest relevance* to near-term Soviet military strength."[30]

SEEKING ACCOMMODATION

National security and *science* are not necessarily antagonistic to one another. In the past, scientists and government leaders in the United States have demonstrated a mutual broad appreciation of the national security concept, including not only military applications and preparations, but also economic, cultural, and other considerations. In his celebrated 1945 report to the president on a program for postwar scientific enterprise, Dr. Vannevar Bush identified an important practical relationship between the nation's security and science: "A broad dissemination of scientific information upon which further advances can readily be made furnishes a sounder foundation for our national security than a policy of restriction which could impede our own progress although imposed in the hope that possible enemies would not catch up with us."[31] This advice, however, with the onset of the Cold War, was disregarded.[32] A 1950 report on the role of the Department of State in the field of national and international science, prepared for the Department by Dr. Lloyd V. Berkner under the sponsorship of the National Academy of Sciences, found "ample evidence that unnecessary restrictions exist on the flow of unclassified scientific and technological information" and, among other relevant conclusions, stated that "these restrictions are dangerous to the progress of United States science, affect adversely the conduct of foreign relations in science, and are therefore damaging to our national security."[33]

Today, for many within the scientific community, national security, in its most meaningful context, is not defined by Soviet military capability alone. Moreover, the prevailing broader view seems to be very close to the Bush perspective. Indeed, reflective of this outlook, the National Academy of Sciences' Panel on Scientific Communication and National Security, after noting that "current proponents of stricter controls advocate a strategy of security through secrecy," indicated that "*security by accomplishment* may have more to offer as a general national strategy." "The long-term security of the United States," said the Panel, "depends in large part upon its economic, technical, scientific, and intellectual vitality, which in turn depends on the vigorous research and development effort that openness helps to nurture."[34]

Achieving national security through secrecy, as opposed to accomplishment, poses an obvious conflict with scientific freedom, one that has been realized in recent government efforts to increase controls on scientific communication and regulate the flow of "sensitive, but unclassified" scientific and technological knowledge. Nonetheless, attorney Harold P. Green, a long-time analyst of scientific freedom issues, finds no "precedent or legal authority that clearly supports the proposition that there is a constitutionally protected right to pursue knowledge or to engage in scientific inquiry." He contends, however, that "such rights are implicit in the First Amendment freedom of speech and press."[35,36] In this regard, the proponents of a restriction, notes Green, carry a heavy burden of proof.[37,38] A particularly strong government interest in a limitation, such as one necessary to protect against a "clear and present danger," must be demonstrated,[39] and the limitation must be structured so that it intrudes upon constitutionally protected utterances to the minimum extent possible.[40,41] As yet, however, no federal court has clearly and directly addressed these considerations relative to constraints on traditional scientific communication that have been extended, in the name of national security, to classrooms, scholarly journals, and professional conferences.[42] Indeed, part of the chilling effect of these constraints appears to be that no one is willing to originate a judicial challenge.

SCIENTIFIC RESPONSIBILITY

Increased national security controls on scientific communication also have implications for scientific responsibility. Undefined as it largely is, scientific responsibility can be envisioned as having many dimensions. Scientists certainly are as morally and legally accountable as any other citizen, but are also thought to bear some additional obligations as a consequence of their professional abilities and knowledge. Indeed, the American Association for the Advancement of Science's Committee on Science in the Promotion of Human Welfare perceived some years ago that, in the post-World War

II era, scientists have pursued a variety of paths in attempting to fulfill their social responsibilities. The panel noted, however, that "no single approach has as yet won the active participation of more than a very small part of the scientific community." Also, "sustained development of these activities" was not evident.[43] Nonetheless, the committee firmly maintained that scientists and their professional organizations "have both a unique competence and a special responsibility" to assist the citizenry with knowledge essential for making informed decisions on science-related public issues.[44]

Maintaining the quality and integrity of the work of the scientific community is also regarded as one of the basic responsibilities of scientists.[45] This might be characterized as mental responsibility, that is, an obligation to use critical examination and review to detect error and fraud. When mental responsibility is extended to technology—in determining the quality of consumer products or the adequacy of a public structure, such as a bridge—its intertwining with moral and legal responsibilities may be more apparent. Such combined responsibilities have been formalized and institutionalized in licensing arrangements for certain science-based professions, such as medicine and engineering, and in independent peer certification of some technology products. Professional codes of ethics reinforce such responsibilities and set standards regarding them.

The secrecy resulting from increased national security controls on scientific communication, however, militates against the fulfillment of such scientific responsibilities. Scientists can be precluded by national security laws from imparting certain information that might otherwise assist the public in discussing and deciding science-based policy issues or that might otherwise maintain the quality and integrity of the work of the scientific community.

SEEKING CLARIFICATION

In the end, what this controversy poses is not merely a limitation on scientific communication, but a curtailment of freedom. It is, therefore, a matter of no small consequence. The situation seemingly would be improved by a better understanding of the national security concept in general, and by a view that favored achieving national security by scientific accomplishment in particular. Furthermore, for various reasons, the prevailing political atmosphere appears to be conducive to some kind of policy change along these lines. Relations between the United States and the Soviet Union are much more relaxed and cooperative; a desire for greater U.S. trade competitiveness overseas remains strong; and, of course, new leadership—a *tabula rasa*—has arrived at the White House.

National security, however, is not the only concept in need of clarification. Scientific freedom, as defined by Dr. John T. Edsall, is an acquired right, that is, a right society is willing to recognize because society stands to benefit in some way. In recognizing scientific freedom as a right, society helps knowledge to accrue, and any knowledge thus accrued could conceivably benefit society. Edsall further suggests that scientists may require some extraordinary freedom allowances in order to fulfill the responsibilities deriving from their professional expertise. "Scientists," he writes, "can claim no special rights, other than those possessed by every citizen, except those necessary to fulfill the responsibilities that arise from the possession of special knowledge and of the insight arising from that knowledge."[46]

Edsall raises several important considerations. For example, how do First Amendment rights, which are "possessed by every citizen," contribute to the concept of scientific freedom? Are there types of science to which society will give more or less freedom because of the different benefits to society that could result? And what special responsibilities are to be fulfilled with the grant of extraordinary freedom?

Some useful contrasts are available if we consider a parallel concept: academic freedom. Walter P. Metzger, a long-time student of academic freedom, has stressed the relationship between faculties and administrations in his work. Suppose that we were to emulate Metzger, and, in discussing the rights and limits of scientific freedom, that we were to stress the relationship between science and society. We might conclude that scientific freedom requires its own arrangements and procedures for detecting and adjudicating offenses, which would not be confined to campus jurisdictions or concerned primarily with threats to employment.[47]

If scientific freedom and scientific responsibility were to be better explained, demarcated, and defined, we could hope that science would be better prepared to confront such challenges as increased national security controls on "sensitive, but unclassified" scientific and technological knowledge or the secrecy assumptions accompanying research sponsored by the federal defense community. As at least an acquired right, scientific freedom is not bestowed by the citizenry merely for the benefit of scientists, but for the advancement of knowledge that may be of some societal benefit. In a democracy, what these benefits, if any, might be is for society, including scientists, to determine.

REFERENCES

1. BECKER, C. L. 1947. Freedom and Responsibility in the American Way of Life: 31. Alfred A. Knopf. New York, NY.
2. BECKER, C. L. 1947. Freedom and Responsibility: 30–31.
3. See, generally, U.S. Office of Technology Assessment. 1986. The Regulatory Environment for Science. U.S. Government Printing Office. Washington, DC.
4. See EHLKE, R.C. & H. C. RELYEA. 1983. The Reagan Administration order on security classification: A critical assessment. Fed. Bar News J. **30:** 91–97.
5. See 35 U.S.C. 181–188. [References in this form should direct the reader to the applicable chapter and section(s) of the U.S. Code – Ed.]
6. 35 U.S.C. 181.

7. 35 U.S.C. 183.
8. 35 U.S.C. 186.
9. See, generally, U.S. Congress. House. Committee on Government Operations. 1980. The Government's Classification of Private Ideas. H. Rept. 96-1540: 1–62. 96th Congress, 2d Session. U.S. Government Printing Office. Washington, DC.
10. U.S. Congress. 1980. Classification of Private Ideas: 21.
11. U.S. Congress. 1980. Classification of Private Ideas: 23.
12. U.S. Congress. 1980. Classification of Private Ideas: 24.
13. See 50 U.S.C. App. 2401–2419.
14. See WADE, N. 1980. Science meetings catch the U.S.-Soviet chill. Science **207**: 1056, 1058.
15. CHALK, R. & S. PAINTER. 1984. National security and scientific communication: Professional society chronology. Committee on Scientific Freedom and Responsibility. American Association for the Advancement of Science. Washington, DC.
16. BOFFEY, P. M. 1982. Censorship action angers scientists. N.Y. Times. September 5: 1, 16.
17. GREENBERG, J. 1982. Remote censoring: DOD blocks symposium papers. Sci. News **122**: 148–149.
18. KOLATA, G. 1982. Export control threat disrupts meeting. Science **217**: 1233–1234.
19. ANON. 1984. AIAA feels chilling effect of DoD censors. Defense Week **5**: 6.
20. See PARK, R. L. 1985. Intimidation leads to self-censorship in science. Bull. At. Sci. **41**: 22–25.
21. Quoted in SHATTUCK, J. 1983. National security a decade after Watergate. Democracy **3**: 57.
22. MOUNTAIN, M. J. 1979. The continuing complexities of technology transfer. Gov. Exec. **11**: 42.
23. See BARNETT, R. W. 1984. Beyond War: Japan's Concept of Comprehensive National Security: 8–10. Pergaman-Brassey's. Washington, DC.
24. ALMOND, G. 1956. Public opinion and national security policy. Public Opinion Q. **20**: 371.
25. HUNTINGTON, S. 1957. The Soldier and the State: 1. Belknap Press. Cambridge, MA.
26. TAYLOR, M. D. 1974. The legitimate claims of national security. Foreign Affairs **52**: 577.
27. CLEM, H. J. & S. L. FALK. 1977. The Environment of National Security: 2. National Defense University Press. Washington, DC.
28. ALLEN, R. V. 1983. Foreign policy and national security: The White House perspective. *In* Agenda '83: A Mandate for Leadership Report. Richard N. Holwill, Ed.: 6. Heritage Foundation. Washington, DC.
29. ULLMAN, R. H. 1983. Redefining security. Int. Security **8**: 133.
30. National Academy of Sciences. Panel on Scientific Communication and National Security. 1982. Scientific Communication and National Security: 53. National Academy Press. Washington, DC.
31. BUSH, V. 1980 (originally 1945). Science—The Endless Frontier: 29. National Science Foundation. Washington, DC.
32. See RELYEA, H. C. 1984. Increased national security controls on scientific communication. Gov. Inf. Q. **1**(2): 187–192.
33. U.S. Department of State. International Science Policy Survey Group. 1950. Science and Foreign Relations: 85. Washington, DC.
34. National Academy of Sciences. 1982. Scientific Communication and National Security: 45.
35. GREEN, H. P. 1977. The boundaries of scientific freedom. Newsl. Sci. Technol. Hum. Values. June: 17.

36. RELYEA, H. C. 1987. The Constitution and American science. *In* Forum Appl. Res. Public Policy. Vol. 2: 106–116.
37. See Organization for a Better Austin v. O'Keefe, 402 U.S. 415 (1970).
38. See New York Times Company v. United States, 403 U.S. 713 (1971).
39. See National Association for the Advancement of Colored People v. Button, 371 U.S. 415 (1963).
40. See Shelton v. Tucker, 364 U.S. 479 (1960).
41. See United States v. Robel, 389 U.S. 258 (1967).
42. See, however, United States v. Edler Industries, Inc., 579 F.2d 516 (9th Cir. 1978).
43. American Association for the Advancement of Science. Committee on Science in the Promotion of Human Welfare. 1960. Science and human welfare. Science **132:** 70.
44. American Association for the Advancement of Science. 1960. Science and Human Welfare: 70–71.
45. EDSALL, J. T. 1975. Scientific Freedom and Responsibility: 8. American Association for the Advancement of Science. Washington, DC.
46. EDSALL, J. T. 1975. Scientific Freedom and Responsibility: 5.
47. METZGER, W. P. 1978. Academic freedom and scientific freedom. Daedalus **107:** 108–109.

Engineering Ethics and the Question of whether to Work on Military Projects

STEPHEN H. UNGER

Department of Computer Science
Columbia University
New York, New York 10027

INTRODUCTION

In her very interesting paper, Deborah Johnson analyzes several attitudes toward professional responsibilities of engineers (and scientists) for the end results of their work.[1] Included in this analysis are comments on my views, as discussed in *Controlling Technology: Ethics and the Responsible Engineer*.[2] I am not quite satisfied with her interpretation of my position, and so I would like to clarify myself and elaborate to some extent. (This effort has benefited from conversations with Johnson about her paper.) I shall also relate some incidents relevant to the general problem, but not connected with the Johnson paper.

THE ROLE OF ENGINEERING ETHICS

Because the paper to which I am responding can be found elsewhere in this volume, it is unnecessary to review Johnson's summary of my previously stated position. But before dealing directly with the issue of military technology, a few words are in order about what I regard to be the general role of engineering ethics.

Along with all civilized people, engineers share an important core of common values and ideals that are recognized (at least in principle) as necessary underpinnings to any decent societal structure. A prime example is respect for human life and welfare. Some of these values are particularly important in the day-to-day work of engineers. The reliance of engineers and scientists on data supplied to them by colleagues makes truthfulness essential to getting the job done. Openness is another value important to progress in engineering and science. Such common values make it possible to formulate a code of engineering ethics that can command wide acceptance among engineers.

At the same time, engineers as well as others vary widely among themselves with respect to political, religious, and cultural backgrounds. Thus, when it comes to the question of what is an appropriate *use* of technology, considerable disagreement may be expected. Beyond sharing the very broad concept that it is wrong to use one's professional talents in such a way as to kill or injure innocent people, or to cause clear harm to the public welfare, engineers—to the same extent as people in general—differ among themselves with respect to what products or projects are commendable or condemnable.

Some differences are on matters of basic values, exemplified by differing positions as to the relative rights of animals and humans. Other differences concern the general behavior of people, for example, the extent to which potential murderers are deterred by the existence of capital punishment. Still other differences involve the relative weights to be assigned to the value of increasing material comforts versus the value of minimizing hazards to life and health. Some have suggested that provisions to resolve such disagreements be incorporated in engineering ethics codes. This seems to me to be inappropriate, and clearly doomed to failure, because there would be no possibility of meaningful acceptance by engineers as a whole. With respect to the question of how an engineer's work may be used (the so-called end-use question), there is no escaping the fact that engineering ethics has only a limited role.

But this does not mean that I believe that the end-use question should not be a professional concern of engineers. On the contrary, the assumed common core of values constrains consideration of professional assignments by the ethical engineer in several ways. There is an obligation to consider both direct and indirect consequences of one's work and to ensure that these are made known to all concerned. (The first rule in my model code concerns this point.)

I believe that engineers have a broad professional obligation to use their talents for morally acceptable purposes. In particular, they should not engage in work that they, on balance, consider detrimental to human health, safety, or welfare. I regret that my formulation of this latter point was apparently not made sufficiently clear in the ethics code proposal cited by Johnson. She implies that engineers with personal moral codes not including respect for human well-being might, consistent with my model professional code, accept assignments harmful to, say, human life. This was certainly not my intent.

WORKING ON MILITARY PROJECTS

How do these ideas relate to the question of the ethics of accepting military-sponsored engineering assignments? On the face of it, a project whose purpose is to develop a lethal military weapon, for example, a surface-to-surface missile, appears to contravene rules against doing harm to humans. Before signing on to such projects, responsible engineers must consider the circumstances very carefully, in the light of their personal value systems and world views. Those who believe in nonviolence as a basic precept will of course reject such assignments out of hand. Others with less absolute views on violence in general might decline on such grounds as opposition to large-scale military forces, or because they believe that the particular weapon involved has no legitimate role in self-defense, or because they distrust those who would be in charge of its possible use. Still other engineers might agree to work on certain military projects that they believe would reduce the likelihood of war, say, by strengthening their country militarily and thereby discouraging potential aggressors. Engineers might also agree to work on weapons systems, perhaps fighter planes, that they believe are most likely to be used in a manner so as to reduce the overall number of human casualties.

The important question is whether the project is likely to lead, on balance, to a net loss or gain in terms of human life, health, and welfare. Those considering devoting their professional talents to such projects have a responsibility to make judgments on their own and to be prepared to defend them. It would be professionally irresponsible to plead ignorance and passively accept the opinions of others. To do so would be to allow oneself to be used in a manner that clearly violates any reasonable definition of professionalism.

Judgments by engineers about the acceptability of projects may also be subject to criticism on the basis of more general moral codes. It would be perfectly appropriate for broader aspects to be discussed in engineering publications and meetings.

Unfortunately, what might appear to be a very minimal position is by no means generally accepted. To my knowledge, no engineering society has a code of ethics including a provision similar to the one under discussion here. A colleague of mine, in a discussion of the Strategic Defense Initiative (SDI), stated that he agreed that it was a fruitless enterprise, but that he did not think there was any way to stop it, and so would be willing to accept SDI funding, which he thought he could put to good use. In this particular case, I suspect that the individual involved, who is a very talented and thoughtful person, may not have been altogether serious; I doubt that he has applied for such funding. But it exemplifies one of the rationalizations used to evade responsibility.

And this responsibility is a heavy one for engineers to bear. The great majority of them are not in business for themselves. They are compelled to choose among limited employment opportunities. A high percentage of engineers are employed, directly or indirectly, on military-related projects, but this use of engineering talent is not a result of wholly free choices engineers have made. Nor is it the result of some inexorable imperative inherent in the nature of technology. Neither the wishes of engineers nor the nature of technology forced the U.S. government to meet the threat of a cutoff in Middle East oil by spending billions of dollars annually on rapid deployment military forces, while virtually phasing out research and development work on alternative energy sources and energy conservation that could make us independent of foreign oil supplies. One cannot, therefore, set overly strict standards for judging whether a working engineer is doing work of positive benefit to humanity.

As a result of their special knowledge, engineers and scientists are more deeply involved in technology-related problems than other citizens. They can help educate others about the issues and can refrain from participating in what they regard as projects that are clearly detrimental to humanity. At the same time, as I shall argue by means of a few personal experiences, it is not always easy to distinguish between good and harmful work. Yet in no way is my treatment of these experiences intended to support an argument that efforts to direct one's work into useful channels are not worth while. If we confined ourselves to endeavors in which success was guaranteed, we would never do anything significant.

LOOKING BACKWARD

About thirty years ago, while in a research group at Bell Laboratories, I initiated and carried out, essentially on my own, a research effort aimed at developing computer methods for doing visual pattern recognition, of the type humans do when they identify hand-produced, irregular-looking alphabetic characters. This led me to the design of one of the first highly parallel processors. I carried this design to the point of detailed logic diagrams, and a technical aid simulated my machine on an IBM 7090 computer. (Realization of such machines in hardware became feasible only after the maturity of integrated circuit technology several decades later.) Simultaneously, I devised a number of ideas for automated pattern recognition, and I was able to implement these ideas on the simulated machine.

My motivation was mainly of an intellectual nature, although I did think that the results might be quite beneficial, for example, to blind people. This work received a good deal of attention. I was disturbed, however, to find that perhaps the greatest source of interest was on the part of the military. Apparently, it was believed that such techniques could be used to guide missiles (then at an early stage of development) to their targets. There was also interest in automating the analysis of photographs taken by spy planes such as the U2. On several occasions I was offered funding from the Department of Defense to continue this work. Being opposed to the militarization of technology, I declined these offers, and, partly as a result of the bad feeling they engendered in me, discontinued my efforts in the area. Subsequent developments indicate how difficult it can be to determine whether the net effect of a project will be harmful or benign.

Such pattern recognition concepts as I had developed were indeed put to use by the military, both for missile guidance (the cruise missile being a prime example) and in analyzing reconnaissance photographs. More recent developments in the area of phased disarmament have put a premium on technology to verify disarmament agreements. Hence the computer analysis of photographs (taken now mostly by satellites) is also a useful tool in dismantling the terrible weapons systems that threaten all of us. Is this particular technology a good or bad thing on balance?

My second tale is about work I did almost immediately after the incident described above. While still at Bell Laboratories, I shifted to more applied engineering work, supervising a group developing software for programming the electronic telephone switching system (ESS), then in the late stages of development. I had no qualms about the end-use of this piece of technology because I believed the telephone system was clearly beneficial. (The junk telephone call was still in the future!)

After nearly two years, my group had produced a very respectable, innovative prototype, and I decided to move to another line of work. Shortly after this decision, I learned that the work we had been doing was considered particularly interesting to a department at Bell Laboratories that was developing a special version of the ESS to be used as a military communications system. They did, in fact, subsequently use our macrocompiler. They may have even used it more than those who wrote the programs for the commercial version. (I believe the military project as a whole, however, was not a big success.) Thus, without knowing it, I had done development work for the military.

Neither of these episodes is of any great consequence, and probably they are not all that unusual. It is all too common for the results of a research project to be used in unforeseen ways that ultimately cause the researcher dismay. One might think that this is less likely for development work, where the applications are far more direct and evident. But my second experience indicates that surprises can occur here too.

One's difficulties in anticipating how one's work will be used is of course no excuse for not trying hard to do good and to avoid doing harm. The existence of ambiguous cases does not justify passivity in clear-cut situations. Although it is sometimes difficult to determine whether or not a project is, on balance, harmful or beneficial, I do not agree with Deborah Johnson's statement that the hard part of being moral is in figuring out what is moral. There is ample evidence that, for engineers, doing what is obviously right can sometimes be extremely costly. There is a pressing need for effective measures to reduce this cost. I believe that a desirable and attainable goal is to achieve a milieu in which engineers, sensitive to the consequences of their work, could engage in open dialogues leading to the intelligent exercise of their consciences without the threat of draconian penalties. Although this would not by itself solve the problem of optimizing the use of technology for the benefit of humanity, it would be a significant step in that direction.

REFERENCES

1. JOHNSON, D. 1989. The social/professional responsibility of engineers. Ann. N.Y. Acad. Sci. This volume.
2. UNGER, S. H. 1982. Controlling Technology: Ethics and the Responsible Engineer. Holt, Rinehart & Winston. New York, NY.

Militarism and the Quality of Life[a]

ALEX C. MICHALOS

Department of Philosophy
University of Guelph
Guelph, Ontario, Canada N1G 2W1

INTRODUCTION

In 1978, the United Nations General Assembly endorsed the Final Document of the First Special Session on Disarmament, which called for the commission of a group of governmental experts to examine the relationships between disarmament and development. Twenty-seven experts were selected from every continent, with Inga Thorsson designated as chairperson. The group's official report was submitted to the Secretary-General of the United Nations in October 1981. The report included nine recommendations, and this essay is a relatively limited attempt to respond to one of them, a quote from which follows below:

> The Group recommends that all governments, but particularly those of the major military powers, should prepare assessments of the nature and magnitude of the short- and long-term economic and social costs attributable to their military preparations, so that the general public can be informed of them.[1]

I will, in the sections that follow, 1) present an overview of some social indicators for Canada and the United States covering the years from about 1963 to 1983; 2) review Canadian federal government expenditures in general for the period from 1974 to 1986; 3) summarize available information on the Canadian arms industry, including production and export figures; 4) present 15 arguments against the Canadian production and export of military arms broadly construed; and 5) provide a brief conclusion.

AN OVERVIEW OF CANADA AND THE UNITED STATES OF AMERICA

Some of the groundwork for the present investigation was prepared in my *North American Social Report*.[2-6] In this work, I compared the quality of

---

[a] This essay is taken from some sections of another essay of the same name that was originally presented at a meeting of the Guelph Chapter of Science for Peace at the University of Guelph, May 24, 1988. The original version was greatly expanded to include responses to 16 arguments opposed to my position, and this expanded version was published in spring 1989 by Science for Peace, University of Toronto.

life of Canadians and Americans in the 1964–74 period in the areas of population structure; mortality, morbidity, and health care; criminal justice; politics; science and technology; education; recreation; natural environment and resources; transportation; communication; housing; economics; religion; and morality and social customs. Broadly speaking, three conclusions were reached. First, "on the basis of an examination of over 135 social indicators and over 1659 indicator values, it seems fair to say that the quality of life in the 1964–74 period was comparatively or relatively higher in Canada than in the United States." Second, "if one looks at the first and last recorded stock values for the usable indicators for each country independently of the other country, . . . [one finds that] both countries improved in more ways than they deteriorated." Third, the responses of national probability samples of Canadians and Americans to over 117 Gallup Poll questions indicate that "the countries were more similar in the 1963–68 period than in the 1969–75 period." In short, "taking the results of my analyses of nonindependent paths [social indicator trends] and opinion poll responses together, it seems fair to say that the countries tended to be or become dissimilar in more ways than they tended to be or become similar."[7]

As for military expenditures, "in the 1965–74 period American military expenditures as a percent of GNP were always two to four times higher than their Canadian counterparts. . . . In the final year the American figure stood at 6% of the GNP, compared to 2% for Canada."[8]

Although I planned to update all my figures to 1984 and later, other projects always got in the way. TABLE 1, however, summarizes the results of comparing the rank-order values of Canada and the United States among 142 countries on 13 indicators for 1983.[9] Canada's rank-orders were preferable to those of the United States on 9 of the 13 indicators. According to Sivard's aggregation procedures, Canada was better off than the United States in "socioeconomic standing" generally and in military expenditures. I suppose these assessments would not have changed much by today. If we may assume that government expenditures for social and economic purposes contribute more to a good quality of life than government expenditures for military purposes, the figures in TABLE 1 suggest that Canadians have been able to make a more favorable trade-off on this score than Americans.

CANADIAN FEDERAL GOVERNMENT EXPENDITURES

TABLE 2 lists the Canadian federal government's total and defense expenditures for the 1974–87 period in millions of current Canadian dollars. In the final year, defense expenditures were estimated to be nearly $10 billion or 8.5% of total government expenditures. Although there has been a steady increase in defense expenditures as a percentage of total expenditures since the 1984 election of a Progressive Conservative government, the 1987 figure is still a bit below the 14-year average of 8.6%. However, according to the

TABLE 1. Ranking of Canada and the United States among 142 Countries on Military and Social Indicators, 1983[a]

Indicator[b]	Ranking Canada	Ranking United States
Military expenditures		
Per capita (−)	26	7
Per soldier (−)	7	4
Per square km (−)	88	23
Average socioeconomic standing (+)	2	3
Gross national product per capita (+)	10	7
Education		
Public expenditures per capita (+)	4	10
School age population per teacher (−)	13	13
Percentage of school age population in school (+)	3	6
Percentage of women in total university enrollment (+)	11	17
Literacy rate (+)	4	4
Health		
Public expenditures per capita (+)	5	10
Population per physician (−)	26	22
Infant mortality rate (−)	9	17
Life expectancy (+)	7	7
Nutrition		
Calorie supply per capita (+)	20	8
Calories as a percentage of requirements (+)	25	14
Percentage of population with safe water (+)	1	14
Total number of indicators for which each country had a preferable ranking[c]		
Negative	4	1
Positive	5	3
Total	9	4

[a] Data are taken from Sivard.[9] See Table 3 and pages 36–37 in Sivard for definitions of specific indicators and their sources.

[b] Plus signs designate positive indicators, and minus signs designate negative indicators.

[c] Excluding the indicator for average socioeconomic standing, and three indicators with tie scores.

Tories' white paper on defense spending, *Challenge and Commitment: A Defence Policy for Canada*, "the Government is committed to a base rate of annual real growth in the defence budget of two per cent per year after inflation, for the [coming 15-year] planning period."[12] At the same time, "after 1986–87, the budget states that operating costs in all federal departments will not be permitted to rise by more than 2 percent in nominal terms each year, which, after inflation, is a real cut of 2 percent."[13] If the defense budget does increase at the projected rate, then in 10 years it will be about 10.5% of total federal government expenditures, which would be roughly its 1972 rate. Putting the 2% real growth rate for defense spending together with the

TABLE 2. Federal Government and Defense Expenditures in Current Millions, 1974–87[a]

Year Ended March 31	Total Expenditures	Defense Expenditures	Percentage of Total Expenditures on Defense
1974	22,839	2232	9.8
1975	29,245	2512	8.6
1976	33,978	2974	8.8
1977	39,011	3371	8.6
1978	42,882	3771	8.8
1979	46,539	4108	8.8
1980	50,416	4391	8.7
1981	58,066	5077	8.7
1982	67,678	6028	8.9
1983	88,521	6938	7.8
1984	96,610	7843	8.1
1985	109,215	8762	8.0
1986	111,227	9094	8.2
1987[b]	116,740	9955	8.5

[a] Data taken from Public Accounts of Canada[10] and Estimates of the Government of Canada.[11]
[b] Values in this row are estimates.

2% real cut rate for all other federal departments, by 1998 the defense budget would be about 12.5% of the total, or roughly what it was in 1970.[14]

TABLE 3 puts defense shares of the federal government's total expenditures in the context of the shares of 16 other functional areas and a residual "others." Because Statistics Canada's accounting procedures are not exactly the same as those of the federal government, there is roughly a percentage point difference between the figures published by the two agencies. So, on average, for example, the defense share according to Statistics Canada was about 7.6% rather than 8.6%. The Statistics Canada figures are preferable for present purposes because the 16 functional areas are more detailed and easier to identify than their counterparts in successive federal budgets.

Inspection of the figures in TABLE 3 shows that the defense share of the total expenditures is typically greater than 13 of the 16 substantive functional areas. Only old age security payments, unemployment insurance payments, and national debt charges typically take bigger slices of the total pie. While the shares of old age security and unemployment insurance payments typically run from one to three percentage points above the defense share, the national debt share typically runs about twice as high as that of defense. At a minimum, what these figures suggest is that Canada's defense expenditures constitute a significant share of the federal government's total expenditures and raise provocative questions regarding the actual versus a more desirable distribution. Because most of this paper consists of specific arguments for less spending on the production and export of military arms broadly construed, more will be said about diverse trade-offs as our discussion proceeds.

TABLE 3. Percentages of Total Expenditures Allocated to Different Functions, 1974–86[a]

Year Ended March 31	Percentage of Expenditures Function[b]																		Total
	1	2	3	4	5	6	7	8	9	10	11	12	13	14	15	16	17	18	
1974	8.8	5.7	7.3	8.0	1.1	12.5	8.9	4.1	3.4	3.8	1.0	0.6	1.8	1.2	7.8	1.5	7.1	15.4	100.0
1975	7.4	5.5	7.1	7.4	1.3	11.1	8.1	5.9	3.1	3.3	0.9	0.7	1.9	1.1	8.7	1.6	7.4	17.5	100.0
1976	7.2	5.1	6.7	7.5	1.6	10.7	9.0	5.3	4.0	3.2	0.8	0.9	2.0	1.4	7.3	1.4	7.7	18.2	100.0
1977	7.8	5.6	6.7	8.0	2.0	10.5	8.9	4.8	4.7	3.3	0.7	1.2	1.9	1.0	8.3	1.4	7.1	16.1	100.0
1978	7.9	5.6	6.4	6.8	2.3	10.6	9.5	4.6	4.0	4.2	0.7	1.1	2.4	1.6	7.6	1.5	7.7	15.5	100.0
1979	8.1	5.5	6.5	7.6	2.6	10.1	9.3	4.1	3.9	4.4	0.8	1.3	1.9	1.4	6.7	1.5	9.3	15.0	100.0
1980	7.7	5.1	5.7	7.3	2.9	11.1	7.2	3.0	3.4	4.2	0.6	1.4	1.8	1.6	7.2	1.7	10.1	18.0	100.0
1981	7.3	5.1	6.3	6.5	3.1	10.9	7.0	2.7	3.3	3.7	0.5	1.5	1.6	1.6	6.5	2.1	10.8	19.5	100.0
1982	7.4	5.1	5.2	6.0	3.2	10.8	7.0	2.6	3.3	3.4	0.4	1.4	1.6	1.4	6.7	1.9	14.2	18.4	100.0
1983	7.2	4.7	3.0	5.0	3.3	10.4	10.8	2.4	3.5	3.1	0.5	1.9	1.7	1.2	6.7	3.1	13.3	18.2	100.0
1984	7.6	4.8	3.1	6.1	3.6	10.2	9.9	2.3	4.1	3.5	0.5	1.6	1.7	1.2	6.4	3.1	12.9	17.4	100.0
1985	7.5	4.7	3.3	6.2	3.8	10.0	9.1	2.1	4.0	3.4	0.4	1.8	1.8	1.0	6.0	2.5	14.7	17.7	100.0
1986	7.4	4.8	3.0	6.1	4.2	10.7	8.9	2.2	3.9	3.4	0.4	1.3	1.8	0.9	5.8	2.8	17.0	15.4	100.0

[a] Data are taken from Statistics Canada.[15]
[b] Column codes: 1: national defense; 2: general government; 3: transportation and communication; 4: health; 5: Canada pension plan; 6: old age security; 7: unemployment insurance; 8: family allowance; 9: assistance to disabled; 10: education; 11: environment; 12: housing; 13: foreign aid and affairs; 14: research establishments; 15: transfers to other levels of government; 16: transfers to own enterprises; 17: debt charges; 18: all other expenditures.

THE CANADIAN ARMS INDUSTRY

Because there is no generally accepted definition of "military arms" broadly or narrowly construed, there is bound to be some controversy over any alleged measured level of production or export of such things. The estimates given here are taken mainly from Treddenick[16] and Regehr.[17] According to Treddenick, "the defence industrial base is that part of the nation's economy providing goods and services required to support military activities."[18] This definition is very broad, but it becomes more useful when it is operationalized by identifying "the defence industrial base in terms of current demands placed on Canadian industry resulting from expenditures for domestic defence procurement and for exports."[18] Refined in this way, the definition suggested by Treddenick comes very close to the one suggested by Regehr: "For purposes of implementing control measures, *the Canadian government should define a military commodity as a commodity purchased by a military force or agency.*"[19] Presumably, both authors would exclude some items like food and housing supplies, and both would include not only weapons but "the support facilities and equipment that make weapons usable."[20,21] Treddenick specifies a "narrow industrial base" within the broader sector, which includes "industries producing specialized military equipment." Operationally, the narrow industrial base includes manufacturers of aircraft, motor vehicles, ships, communications equipment, and some chemicals insofar as the products are sold to military agencies. Applying these rough definitions and some appropriate caveats, he reaches the following conclusions:

> If economic significance means the amount of economic activity generated in the defence industries, then by comparison to total economic activity in Canada, the defence industrial base must be judged to be insignificant. Total defence production accounts for considerably less than one per cent of both gross domestic product and total employment. When the narrow defence industrial base alone is considered, these contributions fall to about one-third of a percentage point in each case. Defence production is also not significant in any single provincial economy. Only in Nova Scotia and New Brunswick does employment generated by defence production approximate one per cent of total provincial employment; in most provinces it is considerably less. Defence production must also be considered insignificant in terms of international trade. Defence exports, net of re-exports, currently account for less than one per cent of total merchandise exports while defence imports, including indirect imports, account for just over two per cent of total merchandise imports. Finally, because of its comparatively low level and because it is difficult to make a theoretical case for its transferability to the civilian sector, defence research and development must also be considered insignificant relative to overall economic activity. . . . The relatively small size of the Canadian defence industrial base makes it extremely difficult to see it as the mainstay of the capitalist system in Canada, in either the Marxian or the Galbraithian sense.[22]

If Treddenick's estimates are about right, in 1986 about $2.5 billion dollars would have been tied up in arms production. Two and a half billion dollars that year would have matched the federal government's expenditures for family allowances and for research and development in the natural sciences. It would have been 6 times the federal government's expenditure on the environment, 13 times the expenditure on research in the social sciences, and 3 times the expenditure on recreation and leisure. It would have just about matched the country's expenses on community colleges, and it would have added another 22,676 new or renovated housing units to those funded by the federal government in 1986.[23,24] Clearly, then, because it would be unreasonable to dismiss all these national expenditures as economically, morally, and politically insignificant, Treddenick's judgment is oversimplified at best.

TABLE 4 and FIGURE 1 provide a longer and more detailed view of Canadian military exports since 1959. The bottom line of TABLE 4 shows that about 73% of our exports have gone to the United States, 14.6% to Europe, and 12% to other places, mainly in the Third World. Although we have just seen that Treddenick would be one of the last people to exaggerate the economic significance of the Canadian arms industry, he remarks that "not only have total exports consistently exceeded domestic demand by large amounts, but exports to the United States alone have done so. . . . This export dependence of the defence industries, and particularly the dependence on a single country, is the outstanding economic feature of the Canadian defence industrial base."[28]

TABLE 4. Canadian Military Exports in Current Millions, 1959-86[a]

Year	Exports in Current Millions[b]			
	United States	Europe	Other	Total
1959-69	2418.8 (78.9)	439.8 (14.3)	207.0 (6.9)	3065.6 (100.0)
1970	226.5 (67.4)	41.2 (12.3)	68.5 (20.3)	336.2 (100.0)
1971	216.3 (64.3)	67.2 (20.0)	53.0 (15.7)	336.5 (100.0)
1972	175.0 (58.3)	73.7 (24.5)	51.7 (17.2)	300.4 (100.0)
1973	198.8 (64.3)	72.8 (72.8)	37.6 (37.6)	309.2 (100.0)
1974	150.0 (53.5)	45.6 (16.3)	84.9 (30.2)	280.5 (100.0)
1975	188.5 (67.1)	58.6 (20.9)	33.7 (12.0)	280.8 (100.0)
1976	191.1 (56.9)	113.1 (33.7)	31.9 (9.4)	336.1 (100.0)
1977	314.1 (56.7)	76.0 (13.7)	163.9 (29.6)	554.0 (100.0)
1978	267.0 (55.1)	129.6 (26.7)	87.9 (18.2)	484.5 (100.0)
1979	367.7 (64.7)	145.6 (25.6)	55.0 (9.7)	568.3 (100.0)
1980	481.7 (66.7)	142.1 (19.7)	97.9 (13.6)	721.7 (100.0)
1981	826.6 (71.8)	149.4 (13.0)	174.8 (15.2)	1150.8 (100.0)
1982	1027.9 (71.7)	157.8 (11.0)	248.4 (17.3)	1434.1 (100.0)
1983	1207.4 (81.5)	128.6 (8.7)	145.2 (9.8)	1481.2 (100.0)
1984	1360.5 (77.6)	243.1 (13.9)	149.8 (8.5)	1753.4 (100.0)
1985	1644.2 (86.8)	154.0 (8.1)	104.5 (5.5)	1902.7 (100.0)
1986	947.0 (68.2)	196.2 (14.1)	244.8 (17.6)	1388.0 (100.0)
Total	12,209.1 (73.2)	2434.4 (14.6)	2040.5 (12.2)	16,684.0 (100.0)

[a] Data are taken from Regehr[25] and Epps.[26]
[b] Percentages are shown in parentheses.

FIGURE 1. Canadian military exports worldwide. Vertical axis: military exports in millions of dollars. Sources: Regehr[27] and Epps.[25]

ARGUMENTS AGAINST THE PRODUCTION AND EXPORT OF ARMS

The following arguments may be offered by Canadians against the production and export of military arms broadly construed.

1. Any resources that are used for the production and export of arms cannot be used for the production and export of such necessities as food, water, clothing, shelter, medical care, and education. Because the latter are more life sustaining than the former, they should not be traded for the former.[12,30]

2. Any resources that are used for the production of arms to be deployed by Canadian forces cannot be used for such necessities as food, water, clothing, shelter, medical care, education, and other social services. According to Werlin, since the Mulroney government came to office in 1984, "transfer payments from Ottawa to the provinces for health care and education have been reduced, unemployment insurance benefits have been cut back, public housing subsidies, environmental protection programs, and support for cultural activities (most importantly the Canadian Broadcasting Corporation) have all been reduced. Increases in the defence budget [averaging 6% a year for the rest of the decade] have quite simply been made by reducing the budget for public services."[31] Thus, in the interest of promoting life-sustaining programs and expenditures, increases in the production of arms should at least be kept below increases in other government programs and expenditures.

3. The availability of arms increases the inclination to resort to and the ability to use violent means to make social and political changes, and de-

creases the inclination to resort to and the ability to use nonviolent means. Again, because the latter are more life sustaining, they should be given priority over the former.[32]

4. Because Canada's biggest trading partner is the United States and about half of our manufacturing industry (not merely our arms industry) consists of American subsidiaries, much of our production and export is designed to American specifications, requirements, and interests. Thus, much of what our arms industry produces and exports is designed to satisfy demands resulting from an American Cold War vision of international Soviet threats to peace. Insofar as this is a biased vision and relatively dangerous for the continued existence of life on the planet, it would be wise to reduce all activities based on this vision, including all activities related to the production and export of arms.[33,34]

5. In order to increase Canada's "capacity to assess and pursue independent and innovative foreign policy and defence options," we ought to decrease opportunities for the United States to influence our decision-making processes. Reductions in the integration of Canadian and American arms industries would significantly decrease such opportunities, and they should therefore be undertaken.[35-37]

6. Because Canada and the United States agreed to have a rough balance of arms trade through the 1963 Defense Production Sharing Arrangements (DPSA), Canada cannot expect long-run economic gains from arms trade with the United States. Hence, continued increases in the production and export of arms must rely on overseas markets. Because Europeans tend to insist on exchanges similar to those of the DPSA, the most promising markets are those of the Third World. If we increase sales of arms to Third World countries, however, we will also increase the carnage of wars, economic dependence, and distorted social development. It has been estimated that as many as 20 million people have died in the 150 wars cited by Regehr.[38] Besides contributing to the massacre, increases in the importation of weapons create foreign exchange shortages that relatively poor countries try to alleviate through greater exploitation of their natural resources for export. Thus, they develop dependent, subsidiary economies incapable of indigenous innovation that might allow them economic independence in the long run. Distorted economies then contribute directly to distorted social development because the greatest burden of government expenditures must be devoted to international debt services rather than to social services.[39] In many ways, as Regehr explains, if Canadians increase their traffic in arms to Third World countries, we will inevitably reproduce some of the debilitating effects on them that trade with the United States has on us.[40]

7. Increases in the production and export of arms to Third World countries would increase the militarization of those countries, not only by increasing the amount of military hardware at their disposal, but by encouraging any tendencies they might have toward forming authoritarian administrations.[41] Thus, because increases in the militarization of Third World coun-

tries tend to undermine their democratic institutions and participative decision making, such increases ought to be resisted.

8. Increases in the production and export of arms will tend to increase the number of people in Canada whose livelihoods depend on militarization. Moreover, it is likely that most (certainly not all) people whose livelihoods depend on militarization will be relatively uncritical of militarization. Thus, because militarization is inconsistent with our democratic institutions, inclinations, and practices, the latter would be undermined by increases in the production and export of arms. Hence, increases in the production and export of arms ought to be resisted in the interest of protecting our own democratic institutions, inclinations, and practices.

9. Because the arms industry tends to be highly specialized and concentrated, labor and material costs tend to be relatively high. Besides, as indicated in argument 11 below, the arms industry tends to be characterized by relatively low productivity. Thus, the combination of these characteristics implies that increases in arms production have inflationary effects as costs outrun productivity gains. Furthermore, Werlin notes that arms production is inflationary because it generates "spendable income without at the same time enlarging the supply of goods available in the market place."[42] Hence, in the interest of reducing inflation, increases in the production and export of arms ought to be resisted.[43-45]

10. It has been estimated that if 1% of the over 50,000 nuclear weapons currently stockpiled by the United States and the Soviet Union were exploded, there would be a nuclear winter of from 6 months to 3 years that most people on the planet could not survive. Because increases in the production and export of arms increases the likelihood of wars, and because any wars could accidentally initiate a global nuclear war leading to a nuclear winter, increases in the production and export of arms ought to be resisted.[46]

11. Because some increases in the Canadian production and export of arms will be connected to the U.S. Strategic Defense Initiative (SDI), and because the latter is seriously defective technologically, militarily, and politically, we should at least resist any strengthening of such connections.[47,48] Tsipis concluded his review of SDI with the pronouncement that it is "voodoo science." According to Nadis, "in October of 1986, a poll found that 98 percent of the members of the [U.S.] National Academy of Sciences in fields most relevant to SDI research believed that SDI could not provide an effective defense of the U.S. civilian population."[49]

12. Because Canada is a relatively small country with a relatively small military establishment and presence in the world, we have the opportunity to initiate changes in our defense policies without creating significant shocks or threats to other countries. New Zealand has denied nuclear-armed American warships the right to enter its ports and has established nuclear-weapons-free zones (NWFZs) covering over 65% of its population. Anderton suggested that a basic premise of such actions is simply that "small nations must take a stand for peace themselves, if they hope to influence big nations."[50]

Thus, Canada might undertake a gradual phasing out of all arms manufacturing that is not directly connected to the particular needs of our own defense policies, assuming that the latter may be specified relatively independently of the policies of the American establishment. As of August 1987, there were over 166 NWFZs in Canada, including the cities of Toronto, Vancouver, Hamilton, and Regina, and the provinces of Manitoba, Ontario, and the Northwest Territories.[51,52] The New Democratic Party has proposed that Canada join Norway, Denmark, Sweden, Finland, Greenland, and Iceland in declaring all our countries a NWFZ, which would be an important building block for an Arctic common security system.[53]

13. The biggest threats to international security in the future are scarcities of raw materials, environmental degradation, declining economic growth, and the severely unequal distribution of the world's wealth. Because increases in the production and export of arms contributes nothing and even decreases resources available to address these problems, in the interest of increasing international security, we ought to reduce expenditures on the former in favor of expenditures on the larger threats.[54-56]

14. Because the need for arms is typically largely in the eye of the beholder, one ought to be skeptical about any alleged need calling for increases in the production and export of arms. Treddenick gives about 10 reasons for defense spending tending to become a bottomless pit, but the following remarks seem to capture a main source of the problem:

> As an economic good satisfying human wants, defence, at least in peacetime, is an abstract concept, one which is technically complex and generally not well understood by the public. It cannot be measured in any objective sense. Whether defence is adequately provided for, or whether the composition of defence spending, including the equipment mix, is appropriate is a matter of perceptions about intentions and relative force sizes, training, tactics, morale and so on. It is therefore impossible to say that there is too much or too little defence spending, or to say that there are too many tanks or too many ships in the same way that it is possible to say that there is too much of other types of goods. The relationship between spending on defence, including how it is spent, and how much defence capability is actually achieved is therefore highly ambiguous. This ambiguity can make defence planning a challenging occupation, but at the same time it provides economic policy makers with an expenditure instrument of a flexibility unmatched by other forms of government expenditure.[57]

15. There is some evidence from a national opinion poll taken in October 1987 that most Canadians would prefer to see less emphasis on a militaristic approach to international security, which would imply less emphasis on the production and export of arms. The survey was sponsored by the North-South Institute (NSI) in Ottawa, and the sponsor's analysts summarized their view of their findings as follows:

> The Canadian public seems to be on a completely different wavelength from its government in what it sees as the main threats to Canadian security and what should

be done about them. In 1987 the government allowed a Defence White Paper to be seen to speak for Canadian security policy, and the Department of National Defence to be seen to shape Canada's views on peace and war. In the year that ended with Mr. Gorbachev in Washington signing the INF [internuclear forces] Treaty, NSI's survey shows most Canadians implicitly rejecting both the Cold War diagnoses and prescriptions of the Defence White Paper tabled by Mr. Beatty [Minister of National Defence] in June.

Canadians themselves have a different and much wider agenda for enhancing international security, including environmental, health, developmental and ethical/political goals. In the maintenance of peace, they seem likely to see Canada's best contribution in more arms control and disarmament efforts, international cooperation, conflict resolution and peacekeeping, rather than in the build-up of arms. Even among various international purposes – quite apart from needs at home – most Canadians resoundingly reject increased defence spending as a priority.[58]

According to Lambert, "polls by Angus Reid and Goldfarb . . . show the majority of Canadians opposed to cruise testing, opposed to the purchase of nuclear submarines, and in favor of making Canada a nuclear weapons free zone."[59]

CONCLUSION

The charter of the United Nations Educational, Scientific, and Cultural Organization reminds us that "since wars begin in the minds of men, it is in the minds of men that the defense of peace must be constructed." I am enough of a feminist to believe that even if the original authors of that sentence understood the term "men" in its generic sense, it has special reference to males. We are the ones typically socialized to be competitive, aggressive, out of touch with our feelings, and all too often arrogantly defensive of our own ignorance. So, I hope that men especially will give serious consideration to this essay.

As I remarked at the beginning, insofar as my arguments are sound, a case should have been made for resisting Canada's proposed increases in the production and export of arms, and for beginning to scale down Canada's current militaristic activities. I have not recommended total disarmament or the gradual phasing out of our military establishment. On the contrary, I have suggested that the military has legitimate national and international functions: routine surveillance, disaster relief, and peacekeeping. We need a defense policy based not on military might but on wisdom, compassion, and diplomacy. As the New Democratic Party of Canada has put it: "Canadian security depends upon a stable international order that recognizes and respects Canadian sovereignty and territory, rather than on Canada's ability to defend itself militarily. Thus, Canada's primary responsibility in its own defence is to contribute to the development of a just international order. The role of the United Nations is central to this process."[60]

ACKNOWLEDGMENTS

I would like to thank the following people for helping me improve the final product: R. G. Good, B. Graf, L. Groake, R. R. Iyer, J. McMurtry, G. Morgan, C. Mitcham, S. S. Nagel, D. C. Poff, J. P. Roos, and J. T. Stevenson.

REFERENCES

1. SANGER, C. 1982. Safe and Sound: Disarmament and Development in the Eighties: 107. Deneau. Ottawa.
2. MICHALOS, A. C. 1980. North American Social Report. Vol. 1: Foundations, Population and Health. D. Reidel. Dordrecht.
3. MICHALOS, A. C. 1980. North American Social Report. Vol. 2: Crime, Justice and Politics. D. Reidel. Dordrecht.
4. MICHALOS, A. C. 1981. North American Social Report. Vol. 3: Science, Education and Recreation. D. Reidel. Dordrecht.
5. MICHALOS, A. C. 1981. North American Social Report. Vol. 4: Environment, Transportation and Housing. D. Reidel. Dordrecht.
6. MICHALOS, A. C. 1982. North American Social Report. Vol. 5: Economics, Religion and Morality. D. Reidel. Dordrecht.
7. MICHALOS, A. C. 1982. Economics, Religion and Morality: 171–174.
8. MICHALOS, A. C. 1980. Crime, Justice and Politics: 176.
9. SIVARD, R. L. 1986. World Military and Social Expenditures 1986. World Priorities. Washington, DC.
10. Public Accounts of Canada. 1986, 1982, 1978.
11. Estimates of the Government of Canada. 1986–1987.
12. Canada, Department of National Defense. 1987. Challenge and Commitment: A Defense Policy for Canada (white paper on defense). Minister of Supply and Services. Ottawa.
13. PRINCE, M. J. 1986. The Mulroney agenda: A right turn for Ottawa? In How Ottawa Spends, 1986–87: Tracking the Tories. M. J. Prince, Ed.: 39. Methuen. Toronto.
14. TREDDENICK, J. M. 1984. The Arms Race and Military Keynesianism. Report No. 3 for the Center for Studies in Defense Resources Management: 21. Royal Military College of Canada. Kingston.
15. Statistics Canada. 1987. Federal Government Finances 1985 (68-211): 41–42, Table 8; Finances 1982: 40–41, Table 12; Finances 1979: 40–41, Table 14; Finances 1976: 44–45, Table 16. Minister of Supply and Services. Ottawa.
16. TREDDENICK, J. M. 1987. The Economic Significance of the Canadian Defense Industrial Base. Report No. 15 for the Center for Studies in Defense Resources Management. Royal Military College of Canada. Kingston.
17. REGEHR, E. 1987. Arms Canada: The Deadly Business of Military Exports. James Lorimer & Co. Toronto.
18. TREDDENICK, J. M. 1987. Economic Significance: 24.
19. REGEHR, E. 1987. Arms Canada: 212.
20. REGEHR, E. 1987. Arms Canada: 70.
21. TREDDENICK, J. M. 1987. Economic Significance: 35–36.
22. TREDDENICK, J. M. 1987. Economic Significance: 50–51.
23. Statistics Canada. 1987. Finances 1985: 41–42.

24. Statistics Canada. 1987. Canada Year Book 1988: 4–5, 7, 11–12. Minister of Supply and Services. Ottawa.
25. REGEHR, E. 1987. Arms Canada: 17.
26. EPPS, K. 1987. Canadian military industrial update. Ploughshares Monitor **8**: 12.
27. REGEHR, E. 1987. Arms Canada: 18.
28. TREDDENICK, J. M. 1987. Economic Significance: 31.
29. REGEHR, E. 1987. Arms Canada: 12.
30. WALLACE-DEERING, K. 1986. The economics of war and peace. In End the Arms Race: Fund Human Needs. T. L. Perry & J. G. Foulks, Eds.: 78. Gordon Soules. West Vancouver.
31. WERLIN, D. L. 1986. Conversion to peaceful production. In End the Arms Race: Fund Human Needs. T. L. Perry & J. G. Foulks, Eds.: 96–97. Gordon Soules. West Vancouver.
32. REGEHR, E. 1987. Arms Canada: xvii, 194–195.
33. REGEHR, E. 1987. Arms Canada: xix, 53, 178, 181–182.
34. DEROO, R. J. 1986. Our war economy and conversion for peace. In End the Arms Race: Fund Human Needs. T. L. Perry & J. G. Foulks, Eds.: 84–87. Gordon Soules. West Vancouver.
35. REGEHR, E. 1987. Arms Canada: xx, 29–30, 179, 185.
36. DEROO, R. J. 1986. Our war economy: 86.
37. WERLIN, D. L. 1986. Conversion to peaceful production: 101.
38. REGEHR, E. 1987. Arms Canada: 11.
39. REGEHR, E. 1987. Arms Canada: 12, 173–174.
40. MICHALOS, A. C. 1982. Economics, Religion and Morality: 58–63.
41. SIVARD, R. L. 1986. World Military and Social Expenditures: 25.
42. WERLIN, D. L. 1986. Conversion to peaceful production: 98.
43. MELMAN, S. 1984. Peace, employment and the economics of permanent war: 2–3. Project Ploughshares Working Paper 84-5.
44. SANGER, C. 1982. Safe and Sound: 41–43, 73–74.
45. REGEHR, E. 1987. Arms Canada: 164–165.
46. PENTZ, M. 1986. To prevent nuclear war and promote nuclear disarmament: It's time for a new look. In End the Arms Race: Fund Human Needs. T. L. Perry and J. G. Foulks, Eds.: 275, 291. Gordon Soules. West Vancouver.
47. TSIPIS, K. 1986. Technical and operational consideration of space-based defensive systems. In End the Arms Race: Fund Human Needs. T. L. Perry & J. G. Foulks, Eds.: 37–46. Gordon Soules. West Vancouver.
48. SIVARD, R. L. 1986. World Military and Social Expenditures: 18.
49. NADIS, S. 1988. After the boycott. Science for the People **20**: 23.
50. ANDERTON, J. P. 1986. Nuclear freedom in one country—how and why: A case study of the development of a nuclear-free policy in New Zealand. In End the Arms Race: Fund Human Needs. T. L. Perry & J. G. Foulks, Eds.: 193. Gordon Soules. West Vancouver.
51. DAVIES, L. & F. MARCHANT. 1986. The special role of municipalities in working for peace. In End the Arms Race: Fund Human Needs. T. L. Perry & J. G. Foulks, Eds.: 239. Gordon Soules. West Vancouver.
52. GAUNDUN, K. 1987. NWFZ: The Canadian scene. Peace Magazine **3**: 11.
53. New Democratic Party of Canada. 1988. Canada's Stake in Common Security. New Democratic Party of Canada. Ottawa.
54. SANGER, C. 1982. Safe and Sound: 29.
55. United Nations, Panel of Eminent Personalities in the Field of Disarmament and Development. 1986. Disarmament and Development. United Nations. New York, NY.

56. CREIGHTON, P. 1987. Cold war heat. Ploughshares Monitor **8**: 5.
57. TREDDENICK, J. M. 1987. Economic Significance: 15–18.
58. North-South Institute. 1988. Fighting different wars: Canadians speak out on foreign policy. Review '87/Outlook '88: 2.
59. LAMBERT, S. 1987. The Canadian peace pledge campaign. Peace Magazine **3**: 24.
60. New Democratic Party of Canada. 1988. Canada's Stake: 50.

Economic Conversion
An Alternative to Military Dependency in the University

JONATHAN FELDMAN

*National Commission for Economic
Conversion and Disarmament
Washington, DC 20003*

Alternatives to the growing reliance of universities, faculties, and students on military research and development can be provided through civilian economic conversion planning. Economic conversion is the political, economic, and technical process for assuring an orderly transformation of labor, machinery, and other resources now being used for military-oriented purposes to alternative civilian uses.[1] Providing faculty, students, and youth outside of academia with economic alternatives to the military requires changes in university, state, and federal budgetary policy. Such alternative budgets require reductions in national military allocations. The deterioration of the national economic infrastructure, seen in the trade deficit and growing obsolescence of the civilian industrial base, provide one rationale for such new allocations. Nationally, conversion has also become critically important in addressing the displacement created by base closures, shutdowns of nuclear production facilities, and military layoffs induced by arms control agreements and a defense procurement slowdown induced by federal budget deficits. The Washington summit bringing together President Reagan and General Secretary Gorbachev at the signing of the Intermediate Nuclear Forces (INF) Treaty symbolized the growing commitment of elites in the United States and the Soviet Union to cut national defense spending as a way to cope with domestic economic problems. A further warning that more cuts could be in the making was former Secretary of Defense Frank C. Carlucci's announcement at the close of 1987 that the military services were instructed to cut $33 billion from the coming year's budget, representing real cuts of 5%.[2-4] One observer has stated that the $33 billion difference was "largely the difference between the former Defense Secretary Caspar W. Weinberger's bloated request and the more modest proposal of Frank Carlucci."[3] Fiscal constraints and the domestic reaction to liberalization in Eastern Europe during the Bush Administration have also led to proposals for substantial defense decreases.

Military budget cuts also threaten to cause displacement in the university. TABLES 1 & 2 show that universities have become increasingly dependent on defense spending in the Reagan era. TABLE 1 shows that many leading uni-

TABLE 1. Defense Department Spending at Twenty Universities, in Thousands of 1982 Dollars[a]

	Fiscal Year 1982	Fiscal Year 1986[b]		Percentage Change
		1986 Dollars	1982 Dollars	
Massachusetts Institute of Technology	216,562	363,925	317,284	47
Johns Hopkins University	235,517	316,831	276,226	17
Stanford University	22,763	37,720	32,886	44
University of Washington, Seattle	10,388	19,981	17,420	68
University of Illinois	8957	15,045	13,117	46
Princeton University	2855	7490	6530	129
University of California, Berkeley	8627	7451	6496	−25
Columbia University	5468	10,387	9056	66
Cornell University	4872	9476	8262	70
University of California, Los Angeles	4546	7451	6496	43
University of Michigan, Ann Arbor	3980	6390	5571	40
Yale University	4264	4467	3895	−9
University of Wisconsin	4611	5393	4702	2
University of Pennsylvania	3790	6119	5335	41
Harvard University	2736	8044	7013	156
Brown University	1942	6125	5340	175
Northwestern University	1916	2934	2558	34
New York University	1896	2960	2581	36
University of Colorado, Boulder	1757	3303	2880	64
University of Connecticut	1241	1600	1395	12

[a] The data for this table have been taken from compilations of statistics issued by the Department of Defense[42] and the Department of Commerce.[43,44]

[b] The conversion between 1986 dollars and 1982 dollars was carried out with a deflator of 1.147.

TABLE 2. Defense Department Support for University Research and Development in Selected Fields, Fiscal Years 1980 and 1986[a]

	Defense Department Support in Millions of 1985 Dollars		Percentage of Federal Support Provided by Defense Department	
	1980	1986	1980	1986
Physics	44.9	59.0	13.8	14.0
Mathematics	21.3	34.5	32.9	35.3
Computer sciences	25.0	59.6	45.8	53.9
Materials sciences	29.4	49.7	38.7	48.4
Aeronautics and astronautics	39.2	36.7	65.6	56.0
Electrical engineering	68.0	83.5	66.5	59.8
Mechanical engineering	19.8	34.8	45.5	48.1

[a] This table has been adapted from one prepared by Holdren and Green.[11]

versities have received more and more defense monies in real terms. TABLE 2 shows that branches of science have also received more of their funds from the military. Such statistics define the critical importance of conversion to helping universities develop alternatives to military support.

But before discussing the potential conversion options of universities, it is worth reviewing the obstacles to weakening the ties between universities and the military economy. Each of the research areas in TABLE 2 corresponds to departments, faculties, and administrators whose dependency on military spending creates constituencies linked to the larger military system. These economic links in some cases extend as far back as World War II, when large military contracts, notably at elite universities, helped consolidate the growing disparity between such schools and the rest of higher education. Many academics who served the military saw Pentagon research as a means to power. For an individual scientist, access to sophisticated equipment could help secure tenure and status. For a defense-dependent university department, power could be gained by a connection to a pipeline of funds that also subsidized the university's budget as a whole (for example, administrative overhead and fixed costs of laboratories).

What might this mean for groups advocating civilian planning in the university? In the past, institutional ties between universities and the military have made it difficult to resist the expansion in Pentagon authority and support for academic budgets. For example, the biology department at the Massachusetts Institute of Technology, distressed by the increasing reliance of their field of research on Pentagon monies, voted to refuse Pentagon funds entirely. The MIT administration, however, being responsible for directing departments highly dependent on military funding, challenged the biology department's decision. The resolution of this dispute has been described as follows:

> To defend the cozy relationship, the administrators threatened to cut off the biology department's in-house funding if the biologists turned down the military money. Squeezed, the biologists reversed their decision and once again allowed Pentagon funding.[5]

In another case, a science writer for the University-Industry Research Program at the University of Wisconsin was dismissed from his job, ostensibly for disclosing his university's efforts to increase its program of biological warfare research.[6] More recently still, an article on Star Wars budget cuts stated that

> some researchers maintain . . . that defense officials might find it beneficial to reduce money for basic research. The universities receiving Star Wars money are fairly well distributed geographically. . . . The Pentagon might let them feel the pinch . . . as a way to generate more political support for a higher budget next year.[4]

In sum, the impact of the military on the universities is far greater than line items representing university research funded by the Department of Defense (DOD).[7-9] The broad scope of military-serving research can be seen in the political influence that the military is able to wield through academic sponsorship. Even those scientists who would rather not take defense money often feel forced to do so. The university system pressures scientists to take money from sources that can annually fund costly research projects. As TABLE 2 indicates, for many branches of science defense monies are likely to provide such funds. Taking such funds becomes the *sine qua non* of life for many academics. Richard De Veaux, an Assistant Professor in the Program in Statistics and Operations Research at Princeton University, has explained that

> it is certainly true that the military provides a substantial amount of funding in the area of statistics and operations research. If I had a strong enough position so that I didn't accept military funding for any basic research the only problems that would cause is that now I've limited myself to one quarter of the available funds. These funds are very competitive to get regardless of the source. . . . [They] are an important aspect in the tenure decision, your ability to provide funds for your research and to support your students . . .[10]

In addition, as professors and students become more dependent on military-serving agencies, there will be a greater tendency for them to fall in step behind the military's objectives in science policy. This result emerges from nothing more intricate than the ordinary effort of faculty and graduate students to think in terms of what research would be of interest to, and hence funded by, the Department of Defense. Such dependency could also shape the political orientations of faculty, affecting what questions they ask and do not ask of their benefactors.

The connection between political orientation and dependency on military research and development was best explained by Under Secretary of Defense Donald Hicks, who, in a recent issue of *Science* (as cited by Holdren and Green[11]), was quoted as saying that he was "not particularly interested in seeing department money going to someplace where an individual is outspoken in his rejection of department aims, even for basic research."[12] Such veiled threats help draw out the "internal regulation" that has governed many academic scientists. Steve Slaby, Professor of Civil Engineering at Princeton University, argued that defense sponsorship of university research has a direct impact on the political stances taken by academic scientists:

> . . . as far as I'm concerned there are strings attached. Maybe very subtle strings, psychological strings. Strings that affect peoples' political outlooks, peoples' political views, peoples' political action. . . . [During] the Vietnam War, some of my colleagues . . . would not sign anti-Vietnam War petitions, yet they would say, "I agree with you Steve, but I can't sign because as you know I get Defense Department research monies."[13]

Similarly, Terry Matilsky, an astrophysicist at Rutgers University, found that such self-regulation has also limited faculty resistance to Star Wars:

> There's a nation-wide petition against Star Wars signed by thousands upon thousands of scientists. . . . You can see the institutions that are highest on the list of signatories. You find that the people who are not accepting [very many] Defense Department contracts are signing it at the rate of 80 to 90 percent. . . . [In] places that do have a lot of military research on campus, such as M.I.T., [the rate] is hover-

ing at around 50 percent. . . . Even within disciplines the same thing is happening. If you look in the Chemistry Department you have a much higher percentage of signatories than say in the Physics Department or Engineering Department.[14]

But the political obedience inherent in military dependency is not just the reflection of what overt political stands military-sponsored scientists take on war and peace. Rather, military dependency determines what kinds of science are taught and what kinds of science are practiced by graduating students. There is a close relationship between science and engineering curriculums and the sponsors of faculty research, as Carl Barus, Professor Emeritus of Engineering at Swathmore College, has noted:

> Professors teach what they know. They write textbooks about what they teach. What they know that is new comes mainly from their own research. It is hardly surprising, then, that military research in the university leads to military-centered undergraduate curricula.

Such links have also had a more organized expression. For example, in 1968, the American Society for Engineering Education published its *Final Report on the Goals of Engineering Education*. A section entitled "The Engineer in Future Society" summarized a report, prepared by the military-linked R.A.N.D. Corporation, in which technical forecasts were made for the years 1984 and 2000. The Society, in its *Final Report*, offered the following comment: "These forecasts suggest that . . . development will continue . . . [in] space programs and [in the] design of more efficient and humane [*sic*] military defense systems." The means by which this comment could become a self-fulfilling prophecy were revealed when the *Final Report* went on to state that it had "attempted to point the way toward the development of engineering education in the decades ahead."[15,16]

If professors teach what they know, and what they know becomes more and more a reflection of military-sponsored university research, then this is bound to affect the career choices of students after they leave the university. They will, as Holdren and Green have noted, "find themselves drawn into careers in military work, not just dissertations, because the narrow and highly applied character of their graduate work leaves them few other choices."[17] The narrowing of the scope of scientific inquiry inherent in the militarization of science can be seen in the decreasing applicability of Pentagon-funded science to civilian needs.[18,19] The economic dependency of scientists on the Pentagon weakens the resistance of the academic community to the military and channels labor into the service of the military. If students and faculty could be provided with economic alternatives to the military, scientists would be freer to address pressing social problems through their work as researchers and as actors in the public realm.

The defense cuts discussed above, together with growing budget deficits, threaten the Strategic Defense Initiative and other large-scale programs that have pumped millions of dollars into science departments across the country.[20] The articulation of alternatives to such defense research programs through conversion plans would provide universities with a concrete option to scrambling over a finite or even shrinking pool of Pentagon funding. In 1970, federal defense cuts[a] led the *New York Times* to declare that a "research emergency" was at hand:

> The Pentagon has never failed to impress the political leaders with its dependence on the creativity of the ablest minds on campus; it would be a devastating commentary on the nation's values if history were to record that the military establishment was the only dependable patron of American research and scholarship, and that when military support was cut off, scientific research withered and died.[21]

This declaration helped define the potential constituency for conversion planning as encompassing the executive officers of universities themselves because ultimately the contradictory ebbs and flows in defense spending have proved an illusory safety net for scientists depending on federal funds for research. A "research emergency" combined with organizing by advocates for peaceful science may push university administrators to use their national lobbying networks for peace rather than war. Administration support for conversion planning is most likely to come from those universities on the periphery of the military economy, that is, from universities other than those garnering the most defense contracts.

Already, several incidents suggest that elements of the national research and development community are interested in pursuing an expansion in civilian research spending. For example, at a recent conference, Columbia University President Michael Sovern told journalists covering the Bush Administration to monitor whether the president calls for a modest shift from applied military research to basic research, a shift that "would be more productive in general and probably enhance our national security."[22] The national defense laboratories at Los Alamos have also planned to place greater emphasis on civilian operations (see APPENDIX).

The successful conversion of defense-dependent university laboratories and the provision of civilian alternatives for scientists depend on a three-tiered strategy with political participation on the university, regional, and national levels. At the university level, past efforts have included an attempt to document work opportunities in alternative energy production. In 1978, the University of California Nuclear Weapons Laboratory Conversion Project studied the possibilities of converting the Lawrence Livermore National Laboratory to alternative production in the energy field. The Project requested and received a detailed computer printout of every employee's job category and salary at the laboratory. After drawing up an inventory of skills and research expertise among university scientists, conversion planners matched them with an alternative agenda for peaceful research. In future projects, the criteria for such research could be developed in consultation with progressive science groups, peace organizations, sympathetic academics, and professional groups.[23] For example, Greg Bischak, an economist with Employ-

---

[a] Department of Defense grants to academic institutions fell from $1.063 billion in fiscal year 1964 to $0.634 billion in fiscal year 1970 and $0.421 billion in fiscal year 1975.

ment Research Associates in Lansing, Michigan, has proposed the development of a "portfolio" of alternative research and development work with an emphasis on basic research (the ideas presented here build on this suggestion).

The formal organization of such conversion planning requires the creation of alternative use committees in defense-dependent universities throughout the country. Such committees could draw on the technical knowledge of scientific laborers and the administrative skills and political connections of administrators; these committees would be divided evenly between administrators and researchers. These two groups would negotiate and plan the development of alternative research programs in science and other military-linked departments throughout the nation.

The development of such programs could lead to the conversion of war research in the schools of physics, engineering, mathematics, and computer science. The schools of biology and agricultural science could explore alternatives to military-sponsored biotechnology and the development of pesticides and pesticide-dependent seeds. Changes in consumer preferences and rising health consciousness have created incentives for a transformation of these research areas. Agricultural industries have been hurt by a shift in demand from beef and canned products to poultry and fresh fruits and vegetables. These changes open up new research areas:

> The research capabilities of the university system can be utilized to conform food commodities to dietary standards and consumer preferences, to invent or develop new processing technologies for foods, and even to create new food products which are amenable to current lifestyles.[24]

An expansion in U.S. programs investigating integrated pest management can also provide alternatives to biotechnology programs. Pesticide toxicity, the evolution of pesticide-resistant strains, and the pollution of aquifers and water supplies by pesticides, plant nutrients, and soil additives each require research into alternative methods of commodity production that would reduce dependence on agrochemicals.[25] Programs examining solutions for the acquired immune deficiency syndrome and other health hazards can also provide an alternative for biologists and medical researchers.[26]

An alternative to scientific and engineering resources devoted to military production can be found in the expansion of research on renewable energy resources such as solar power, hydroelectricity, cogeneration, and alcohol fuels from biomass. Such energy research, together with alternative medical and pest management research, can be shared with Third World nations seeking solutions to their own scientific problems. Japan is developing a comprehensive alternative to the Strategic Defense Initiative, America's major high-technology program, through the Human Frontiers Science Program. Although the Japanese have emphasized "big science," which means there may be some risk of overcentralization, their program is directed toward studying "energy conversion from light (photosynthesis) and other sources, to electrical, chemical and kinetic energy."[26]

How can students and faculty gain the political leverage needed to bring about the conversion of universities? How can universities, which are pushed to (or are already sympathetic with) conversion, finance planning and alternative research programs? The conversion of universities depends on the growth of broad-based coalitions. The economic depletion of the United States, as evident in a decaying infrastructure and productive base, has made whole classes of professions and social groups potential beneficiaries of a national economic program to rebuild decaying civilian industries and convert military facilities to civilian uses. The industrial decline of the United States, our lack of industrial competence or competitiveness, is partially rooted in the diversion of scientific and engineering talents to the military. University scientists and engineers can play a pivotal role in rebuilding the U.S. economy in projects as diverse as prefabricated housing for the homeless and high-speed, energy-efficient mass transit. They can also help retrain their counterparts in the defense sector whose socialization to the patterns of military-serving research have created a trained incapacity for alternative civilian work.[27,28]

A program of economic revitalization and conversion tapping the resources of the university could also benefit communities that are often excluded from academic resources or marginalized by university programs. Peace, environmental, and labor groups, and associations representing women, people of color, the working class, and the poor, all have a stake in what priorities are set for the universities and the nation's research and development programs. These groups can apply (and have applied) pressure on the universities to devote more attention to frequently neglected areas of inquiry: women's studies, black and Afro-American studies, and research making disarmament a central concern. Coalitions can begin to push universities to develop alternative budgets that would provide less support for military research and more support for peaceful programs.[29]

The conversion of universities also depends on the participation of campus-based and progressive coalitions in local and national efforts to convert the economy. If industries make more civilian products and less military ones, markets will emerge to support civilian research and development. Furthermore, professional associations and business interests tied to alternative research and development policies could become part of university, local, or national conversion movements.[26] Although basic research, even science designed for peaceful uses, would be (as it usually is) capable of being exploited by the military,[30,31] a converted economy would create barriers for technology transfers from civilian to military uses. For example, expanded funding for civilian research, contracted military budgets for military-sponsored research, and general cutbacks in DOD and DOE programs would lessen the demand for military-linked research and development performed by universities.

The economic literature on the military economy and the growing dependence of domestic institutions on Pentagon capital indicates that many universities, like prime military contractors, will resist conversion unless they are forced by legal and political means.[32-34] Steps toward a legal mechanism to apply such constraints locally has been developed in the City of Berkeley. In November 1986, voters there overwhelmingly passed the Nuclear-Free Berkeley Act. A specific clause in the act mentions legal constraints on universities: "No person, corporation, university, laboratory, institution or other entity shall, within the City of Berkeley, knowingly engage in work for nuclear weapons." The act requires the cessation of nuclear weapons work within city limits and the divestiture of city funds from businesses that engage in nuclear weapons work. The act also promotes "educational activities . . . to advance public awareness and understanding" about the dangers of nuclear weapons.[35-37]

The necessary planning for economic conversion is defined by a 1989 bill in the Congress, introduced into the House of Representatives by New York Representative Ted Weiss, which had more than 50 more House sponsors in the last congressional session. The Defense Economic Adjustment Act (or HR 101) establishes "alternative use" planning committees at every military base and military industrial facility including university research laboratories and "think tanks" which receive defense contracts.

The Weiss Bill would require universities and other contractors (as a condition of receiving defense funds) to pay into an economic adjustment fund an amount equal to one and one fourth percent per year of the value of the contractor's gross revenue on defense sales.[38-40] Thus, in response to university claims that they can not afford conversion planning, the Weiss Bill permits a process whereby university conversion is self-financing through defense contracts.*b*

Universities can stand together to mobilize their collective resources for the civilian management of our nation's research and development resources. This may strike many as a naive and utopian proposal. History, however, has given us a fortunate example of such an attempt. In the mid to late 1970s, twelve southwestern universities attempted to create a consortium to manage the Los Alamos Scientific Laboratory, thereby displacing what had become the permanent manager, the University of California. The twelve schools meant to break a pattern whereby "competitive proposals had never been sought in selecting a contractor to manage a major weapons laboratory."[41] In recent years, universities have lobbied for increased military funding and have participated in such quasi-lobbying groups as the University–Department of Defense Forum, which includes several university presidents and Defense Department officials. As the defense budget is reduced, however, universities may be looking to mobilize public support for national civilian alternatives, alternatives that have a more powerful claim to legitimacy than a program of military expenditures discredited by waste, fraud, and incompetence—even at producing military hardware. By examining the civilian alternatives to defense expenditures on a department-by-department basis, the aca-

---

b The bill also requires a one-year prenotification of plans to cut back or terminate a defense contract or military base. Planning assistance is provided together with income support and retraining programs for communities and workers while a conversion is underway.

demic community can help draw up a portfolio of research and development priorities that could advance public infrastructure investments. This portfolio would define the macroeconomic parameters of a new national civilian economy. It would reflect and advance the national interest that has increasingly been defined by the majority of the population as the advance of national security through economic rather than military development.

REFERENCES

1. FELDMAN, J. 1988. An Introduction to Economic Conversion. National Commission for Economic Conversion and Disarmament. Washington, DC.
2. HALLORAN, R. 1987. Carlucci orders $33 billion in cuts for the armed forces. N.Y. Times. December 5: 1, 35.
3. BISCHAK, G. 1987. Pentagon legerdemain (letter). N.Y. Times. December 22: A22.
4. CORDES, C. 1987. Many scientists welcome the reluctance of Congress to back large increases for "Star Wars" research. Chron. Higher Educ. December 16: A17–A18.
5. SHULMAN, S. 1987. Poisons from the Pentagon. Progressive. November: 18.
6. JANNACCIO, R. 1987. The University of Wisconsin hatches toxins. Progressive. November: 20.
7. KLARE, M. T. 1969. The University-Military Complex: A Directory and Related Documents. North American Congress on Latin America. New York, NY.
8. MELMAN, S. 1970. Pentagon Capitalism. McGraw-Hill. New York, NY.
9. NOBLE, D. 1984. America by Design. Oxford University Press. New York, NY.
10. DEVEAUX, R. 1987. Professors and the Pentagon (as quoted in the April 16 airing of Currents, a WNET-TV documentary series). Public Broadcasting Service. New York, NY.
11. HOLDREN, J. P. & F. B. GREEN. 1986. Military spending, the SDI, and government support of research and development: Effects on the economy and the health of American science. F.A.S. Public Interest Rep. 39(7): 15.
12. SMITH, R. J. 1986. Pentagon's R&D chief roils in the waters. Science 232(4749): 444.
13. SLABY, S. 1987. Professors and the Pentagon (as quoted in the April 16 airing of Currents, a WNET-TV documentary series). Public Broadcasting Service. New York, NY.
14. MATILSKY, T. 1987. Professors and the Pentagon (as quoted in the April 16 airing of Currents, a WNET-TV documentary series). Public Broadcasting Service. New York, NY.
15. BARUS, C. 1987. Military influence on the electrical engineering curriculum since World War II. IEEE Technol. Sci. Mag. 6(2): 5.
16. BALABANIAN, N. 1969. The essential focus of engineering education: The individual student. IEEE Trans. Educ. E-12: 1–3.
17. HOLDREN, J. P. & F. B. GREEN. 1986. F.A.S. Public Interest Rep. 39(7): 14.
18. ROARK, A. C. 1987. Research: Emphasis is military. L.A. Times. April 12: 1, 3.
19. Department of Defense. 1987. Department of Defense Report on the University's Role in Defense Research and Development (for the Committees on Appropriations). Washington, DC.
20. On the SDI's effects on universities, see NIMROODY, R. 1988. Star Wars: The Economic Fallout. Chapter 6. Ballinger. Cambridge, MA.
21. Editorial. 1970. Research emergency. N.Y. Times. April 30: 34.
22. ANON. 1989. Sovern asks watch on Bush pledge. Columbia University Record 14(16): 1, 6.

23. For a description of the University of California conversion project and an outline on conversion planning basics, see SCHUTT, R. 1984. Economic conversion planning. *In* The Military in Your Backyard: 115–126. Center for Economic Conversion. Mountain View, CA.
24. California State Senate Select Committee on Long-Range Policy Planning (Senator J. Garamendi, Chair). 1986. California and the 21st Century: Foundations for a Competitive Society. Vol. 1: 51. Joint Publications. Sacramento, CA.
25. Senate Select Committee. 1986. California and the 21st Century. Vol. 1: 60–61.
26. PIANTA, M. 1988. High-technology programmes: For the military or the economy? Bull. Peace Proposals 19(1): 53–79.
27. For an elaboration of the argument that the military co-opts scarce scientific talent, see DUMAS, L. J. 1986. The Overburdened Economy. University of California Press. Berkeley, CA.
28. On the problems of engineers in military-serving firms, see MELMAN, S. 1987. Profits without Production. University of Pennsylvania Press. Philadelphia, PA.
29. For a description of the work carried out by the National Coalition for Universities in the Public Interest, see KRINSKY, R. 1988. Swords and sheepskins: Militarization of education in the United States and prospects of its conversion. Bull. Peace Proposals 19(1): 47–48.
30. WEIZENBAUM, J. 1986. Not without us. Sci. People. November/December: 18–20, 37.
31. ARISTOVE, N., C. REGEN & E. SMITH. 1986. Ethical dilemmas: Between a rock and a hard place. Sci. People. November/December: 8–9, 11, 13.
32. MELMAN, S. 1974. The Permanent War Economy. Simon & Schuster. New York, NY.
33. GANSLER, J. 1980. The Defense Industry. MIT Press. Cambridge, MA.
34. KAPLAN, F. 1987. Defense profits are double commercial profits, study says. Boston Globe. May 13: 9.
35. SKINNER, N. 1987. Implementation of Measure K, the Nuclear Free Berkeley Act (City Councilor's Memo). January 12.
36. City of Berkeley. 1986. The Nuclear Free Berkeley Act. Ordinance No. 5784-N.S. December 19.
37. For a discussion of the prospects for economic conversion on the regional level, see FELDMAN, J. 1988. Converting the military economy through the local state: Proposals for economic conversion in Massachusetts. Bull. Peace Proposals 19(1): 99–116.
38. DUMAS, L. J. 1984. Making peace possible: The legislative approach to economic conversion. *In* Economic Conversion: Revitalizing America's Economy. S. Gordon & D. McFadden, Eds.: 67–85. Ballinger. Cambridge, MA.
39. Defense Economic Adjustment Act. 1989. H.R. 101. 101st Congress, 1st Session. January 3.
40. FELDMAN, J., R. KRINSKY & S. MELMAN. 1988. Criteria for Economic Conversion Legislation. 1st edit. National Commission for Economic Conversion and Disarmament. Washington, DC.
41. For a fuller description of this effort and related citations, see FELDMAN, J. 1989. Universities in the Business of Repression. South End Press. Boston, MA. See also BJORK, L. G. 1983. Organizational environment and the development of a research university. Ph.D. Thesis. University of New Mexico. Albuquerque, NM.
42. U.S. Department of Defense. 1986. Educational and Nonprofit Institutions Receiving Prime Contract Awards for R.D.T. & E. (fiscal years 1982 and 1986). Directorate for Information Operations and Reports. Washington, DC.
43. U.S. Department of Commerce. 1986. Survey of Current Business. Vol. 66. No. 7.
44. U.S. Department of Commerce. 1987. Survey of Current Business. Vol. 67. No. 5.

APPENDIX

Research in Support of Technological Competitiveness[c]

JOHN J. WHETTEN, *Associate Director*
MICHAEL G. STEVENSON, *Deputy Associate Director*

*Energy and Research Applications Directorate
Los Alamos National Laboratory
Los Alamos, New Mexico 87545*

Los Alamos National Laboratory has grown over the last four decades into a large, multiprogram national laboratory. We are operated by the University of California for the Department of Energy and currently employ nearly 8000 technical and support staff with an annual budget of about $875M operating plus an additional $100M capital equipment and construction. Including contractor employees, nearly 12,000 people work at the Laboratory in northern New Mexico.

Our primary mission today is still nuclear weapons technology, and this Department of Energy program [accounts for] about 50% of our total funding for FY 1988. We also conduct nonnuclear defense work for the Department of Defense, and this work accounts for about 27% of our funding. Nondefense research and technology for the DOE and other sponsors makes up the other 23%.

At Los Alamos, we have been able to advance the frontiers of science and developing technologies with end applications in both the defense and civilian sectors. Research and development in nondefense areas are vital to the scientific excellence of a defense-oriented laboratory. We have made significant contributions to energy technologies, both nuclear and nonnuclear. We are now contributing to health and environmental research [and] space science and technology, and are moving rapidly to enhance the transfer of our technology developments to the private sector.

The Energy and Research Applications Directorate at Los Alamos, which we represent, has the management responsibility for all energy and nondefense space programs at the Laboratory and for developing new programs in nondefense areas such as environmental research and development and advanced space technology development. We also manage the Laboratory's programs in verification and safeguards research and development and are responsible for the Laboratory's technology transfer activities. The Directorate has a total of over 1000 people, most in four technical divisions, Earth and Space Sciences (ESS), Nuclear Technology and Engineering (N), Controlled Thermonuclear Research (CTR), and Mechanical and Electronic Engineering (MEE). Finally, our Directorate contains the Los Alamos branch of

---

[c] Summary of a briefing—January 28, 1988.

the University of California's Institute for Geophysics and Planetary Physics (IGPP). The programs we manage for the Laboratory account for about $150M.

These resources and capabilities will continue to contribute in major ways to energy and related technology areas and in other areas more closely associated with the nation's security, such as arms control and verification technology development. We believe we need an expanded and newly focused program for energy security in this country. The most effective use of federal resources is to better balance the program between technologies having reasonably near-term potential such as enhanced oil recovery research and clean coal technology, with those having midterm payoff possibilities such as passively safe nuclear reactors, and those with long-term potential such as fusion energy.

We also must focus on what we feel is a most significant growing role for the Laboratory—research in support of technological competitiveness. We believe strongly that we and our sister DOE laboratories can and must do much more to help economic development in this nation. This will take new approaches and new alliances. We must solicit more input from the private sector, and develop new partnerships with [it]. Technology transfer from the DOE laboratories, although with many successes to its credit, must become more aggressive, more deliberate, and more open. We must make the laboratories more user-friendly by revising current restrictive policies on contracting with the private sector and intellectual property rights. We must also bring the nation's universities into new cooperative arrangements both with the laboratories in joint research programs and with the laboratories and industry in cooperative enabling technology development programs, such as being proposed in the high-temperature superconductivity area.

In our Directorate, we have several major ongoing efforts in research and technology development which can support increased national security, including energy and economic security, and technological competitiveness. These include our recent efforts to explore with industry the establishment of cooperative programs to develop enabling technologies for commercial high-temperature superconductor applications. In the energy technology arena, our hot dry rock geothermal energy program, our fission and fusion energy programs, the Central American energy and minerals program, and our fuel cells work are all making major contributions toward enhanced energy security. Finally, we have a number of initiatives proceeding to establish new joint programs with universities, including a joint proposal with two other University of California IGPP branches to establish an NSF Center for Computational Geophysics.

Looking to the future and to how we can contribute in a larger way to [solving] emerging national problems, we expect to see and are working hard to establish a growing role for the Laboratory in environmental research and development, particularly in development of environmental cleanup technologies as related to hazardous waste problems in the defense community. We are also looking at how some of the advanced technologies developed in the hot dry rock geothermal program and in other programs can be applied to enhanced oil recovery. We are beginning to establish much stronger interactions with the private sector oil and gas R&D community and are finding that there indeed are several Laboratory-developed research and technology products of considerable interest. Beyond the energy area, our Directorate is looking forward to making new and expanded contributions in nuclear testing treaty verification and to examining applications of existing technologies, largely derived from nuclear materials detection and safeguards programs, to INF and potential strategic arms reduction treaty verification problems.

Finally, we are strengthening our Laboratory's technology transfer activities. We will in the near future double the level of effort in our Industrial Applications Office, and we are working to expand the technology transfer activities in all of our lead technology development divisions. We are exploring with the Department of Energy ways to improve policies and practices on contracting with the private sector and on intellectual property rights. In short, to paraphrase President Reagan's statement of February 17, 1987, we are working hard to assure that we maximize our capability "to aid in making American products and technology better and more competitive."

PART IV. RECONSIDERATIONS OF THE STRATEGIC DEFENSE INITIATIVE

The Ethical Imperative for Limited Strategic Defense

HARRISON H. SCHMITT

P.O. Box 14338
Albuquerque, New Mexico 87191

INTRODUCTION

As a U.S. senator in December of 1980, I met privately with President-elect Ronald Reagan to discuss a broad range of science and technology issues facing the new administration. Prefacing a question about the potential of lasers as a defense against missiles, the newly elected President said he did not believe that the continued production and deployment of weapons of mass destruction could preserve the peace indefinitely. He further stated that although we must continue the policy of retaliatory deterrence for the time being, we also must search for defensive alternatives.

Those were sound instincts then—they are sound instincts now.

Strategic defense issues and opportunities before the world carry extraordinary weight in considerations of the future of humankind and freedom. We clearly have no ethical choice but to do everything possible through technological and diplomatic ingenuity to protect humanity from the threat of nuclear, chemical, and biological weapons. Protection against mass destruction or mass intimidation must ultimately include, if humanly and technologically possible, making such weapons, in Reagan's original terms, "impotent and obsolete."[1] What technology has unleashed, technology must attempt to harness. There is no other ethical alternative.

In this spirit, President Reagan formally suggested in March of 1983 that this nation should begin to apply its scientific and technological talents to the search for defensive alternatives to the strategic policy of the 1960s and 1970s, that is, Mutual Assured Destruction, or MAD. This call for a Strategic Defense Initiative (SDI) probably should have received more thoughtful consideration from some critics that it did when put forth by Reagan. Indeed, many seemed to have reacted more to the proposer than to the proposal. Some proof of this comes in an article which appeared in the *New Yorker* of November 19, 1979, and which was written by a good friend, Daniel Patrick Moynihan of New York.

In that well-received essay, "Reflections on the SALT Process,"[2] Senator Moynihan makes almost identically the same point as President Reagan when he quotes favorably the Soviet dissident, Andrei Sakharov, who in turn was quoting the American physicist Freeman Dyson:

Somewhere between the gospel of nonviolence and the strategy of Mutual Assured Destruction there must be a middle ground that allows killing in self-defense but forbids the purposeless massacre of innocents.

The debate over strategic defense during the succeeding decade might be summarized by the following question: Are strategic defenses possible that do not create more serious problems than they solve?

If the answer to this question is yes, the pragmatic but still humanistic goal for strategic defense must be to reduce the value of offensive strategic weapons to the point where such weapons no longer provide an effective or affordable first-strike capability. With the realization of this goal, offensive strategic weapons would, for all practical purposes, become impotent and obsolete, whether or not a theoretically leak-proof system existed.

As an important side note, this discussion relates to strategic weapons with delivery systems that traverse space or the upper atmosphere. Nuclear, chemical, or biological weapons that can be delivered clandestinely or via cruise missiles require different defenses and, potentially, represent a far more difficult problem for the international community.

AN INTERNATIONAL LIMITED STRATEGIC DEFENSE

As many point out, danger exists in any one nation's apparent if not real success at developing and deploying strategic defenses. For example, the United States already faces a significant level of such danger. The Soviets have created asymmetries by deploying missile defenses around Moscow, by aggressively exploiting battle management loopholes in the Antiballistic Missile (ABM) Treaty, and by instituting extensive civil defense preparations. Hypothetically, a less successful adversary ultimately may feel obligated to act, to "use it or lose it," or see its deterrent force become useless. The danger that could arise from one or another adversary's greater success in developing a defensive system could be eliminated by ensuring that each potential adversary's strategic defenses grew in concert with those of others. How better to do this than through the institutionalized sharing of technology?

In this regard, insight and opportunity lie behind Ronald Reagan's "Why not?" answer to the 1984 presidential debate question about sharing strategic defense technology.[3] It opens the way to breaking the continuing domestic and international impasse on the advisability of developing defensive strategic systems in preference to offensive strategic systems.

Domestically, the psychological attachment we have to the ABM Treaty constitutes the greatest political hurdle faced in the development of a strategic defense system. On the other hand, willingness to share strategic defense technology could be the basis for negotiating politically acceptable amendments to this treaty that would unambiguously permit an internationally managed limited strategic defense system. A limited system, jointly operated by the

United States, the Soviet Union, and other contributing nations, could have as its sole purpose the protection of all peoples against accidental or terrorist missile launches. With this start, even broader international arrangements ultimately should be possible.

Internationally, current aspects of the worldwide political environment enhance the prospects for multilateral strategic defense. First and foremost, multilateral agreements to cooperate in strategic defense research exist between the United States and several of its defense partners. Second, Reagan and Gorbachev have created an apparently improved climate for innovative discussion between the West and the Soviet Union. Third, concerns over the proliferation of nuclear, chemical, and biological weapons, and their means of delivery, have been heightened by the use of long-range rockets in the Iran-Iraq War, by steps taken in Libya, Iran, and Iraq toward the production of chemical weapons, and by the increasing potential availability of nuclear weapons to more and more nations, and to terrorists. At the same time, the United States and (apparently) the Soviet Union have begun to destroy current stocks of chemical weapons, and they appear to be moving toward the cessation of new production.

The technological basis clearly exists for effective worldwide defense against a small number of space or atmospheric delivery systems for strategic weapons,[4] whatever the potential may be for a broad-based strategic defense system. In addition, there are international precedents and experience in managing complex technological endeavors of global impact, particularly with respect to international telecommunications satellites, world weather monitoring and forecasting, Antarctic environmental research, and the destruction of the intermediate nuclear forces of the Soviet Union and the United States.

The suggestion of an internationally managed system for limited strategic defense was first presented at the Eleventh National Security Affairs Conference at the National Defense University in December of 1984. The possibility of a limited, United States-only system—an accidental launch protection system (ALPS)—subsequently received national attention as a suggestion by Senator Sam Nunn[5] and as a broadly supported but unsuccessful Senate amendment to the Defense Authorization Bill for fiscal year 1989. Internationalizing a limited system, however, not only solves the problem posed by unequal rates of technological progress and lays the foundation for sharing future developments, but logically extends two centuries of evolution in the formulation of U.S. defense and foreign policy.

EVOLUTION OF U.S. DEFENSE AND FOREIGN POLICY

The established consensus in the defense and foreign policy of the United States has evolved dynamically over 200 years. Although at each point in history, this consensus has been interpreted rather dogmatically by particular interests at different times, its spatial bounds have grown from isolationism to outer space, and its technological bounds have grown from the range of an eighteenth century cannon to that of a twentieth century intercontinental missile. This evolution, however, forms a continuum with modern consideration of a limited strategic defense system and, indeed, the placement of such a system under international control.

George Washington, in his Farewell Address in 1796, said that "it is our true policy to steer clear of permanent alliances with any portion of the foreign world."[6] For over a century and a half, this statement set the basis for the country's reluctance in making commitments in foreign policy, a reluctance that was easily maintained while national attention was focused on settling a continent. On the other hand, James Madison was convinced as early as 1812 that "acts hostile to the United States as an independent and neutral nation"[7] required more than words in response.

The expansion of the sphere of world interest of the United States took another step in 1823 with James Monroe's statement about Latin America, later known as the Monroe Doctrine: "The American continents, by the free and independent condition which they have assumed and maintain, are henceforth not to be considered as subjects for future colonization by any European powers . . ."[8] The global expansion of U.S. interests to affairs beyond the western hemisphere came in 1904 when Theodore Roosevelt embraced Mahan's extension of the Monroe Doctrine, saying at one point, "If we are not to have . . . a navy, then we should be manly enough to say that we intend . . . not to try to exercise that influence in foreign affairs which comes only to the just man armed who wishes to keep the peace."[9]

An even broader interpretation of U.S. interests and responsibilities than that espoused by Theodore Roosevelt came permanently with the nation's entry into World War II. Although foreshadowed by Woodrow Wilson's post-World War I advocacy of the League of Nations, the 1941 declarations of Franklin Roosevelt and Winston Churchill in the Atlantic Charter[10] laid the foundation for U.S. participation in the United Nations in contrast to its post-World War I rejection of the League of Nations. Further, a truly global defense and foreign policy was the direct consequence of the successful outcome of U.S. involvement in World War II and the simultaneous emergence of the global ambitions of the Soviet Union.

An additional outgrowth of World War II and Soviet expansionism was the commitment of the United States to multilateral defense alliances, such as the North Atlantic Treaty Organization (NATO). Harry Truman's leadership and the Senate's 1948 Vandenberg Resolution set as a bipartisan objective the "progressive development of regional and other collective arrangements for individual and collective self-defense . . ."[11]

The technological advancement of the means of transportation, communications, and war have always had a controlling but backstage influence in the evolution of the accepted boundaries of U.S. defense and foreign policy. Airpower and the introduction of nuclear weapons and the global means for their delivery, however, brought science and technology to the forefront of policy determination during and after World War II. The first clear acceptance by the United States of the moral imperative of controlling the consequences of weapons technology came with the signing and ratification of the Nuclear Test Ban Treaty in 1963. John Kennedy announced the signing of this treaty between Great Britain, the Soviet Union, and the United States as follows: "For the first time, an agreement has been reached on bringing the forces of nuclear destruction under international control."[12] With the Test Ban Treaty and its progeny, the 1972 ABM Treaty, control was limited, however, by Soviet insistence that compliance with these treaties be monitored by remote means.

Kennedy also advocated direct U.S. cooperation in the international application of new space technology when he delivered a policy statement on communications satellites in 1961.[13] In this statement, he invited "all nations to participate in a communications satellite system in the interest of world peace and closer brotherhood among peoples of the world." Practical cooperation in this arena came in 1964 with successful U.S. advocacy and assistance in the creation of the International Telecommunications Satellite (INTELSAT) organization.[14] Not only did the United States transfer its telecommunications satellite technology to an international organization, but it subordinated its potential for unilateral control to the actual majority control of other users. Although INTELSAT did not include participation by Soviet Block nations, they are included in INTELSAT's progeny, the International Maritime Satellite (INMARSAT) organization, established in 1978.

The most recent historical step in the evolution of U.S. defense and foreign policy toward cooperative management of global weapons technology came in 1987–1988 with Ronald Reagan's advocacy and the Senate's ratification of the Intermediate Nuclear Forces (INF) Treaty. Under this treaty, the Soviet Union and the United States are cooperating for the first time in the destruction of a small class of nuclear weapons delivery systems under a limited program of on-site inspection and verification.

This view of the evolution of U.S. defense and foreign policy from isolationism to global cooperation indicates that U.S. participation in the international management of a limited strategic defense system may not be as far-fetched as it might first appear. Such a system could readily build upon the combined management and shared control precedents set by NATO, INTELSAT, and the INF Treaty.

The internationalization of strategic defense does not require that sovereignty over one's own defense be given up; it does not require that technological opportunities to enhance national security be abandoned; it merely requires that some sovereignty and opportunity be shared. We could emulate the vision behind Ronald Reagan's "Why not?" response to the question of whether he would share strategic defense technology with the Soviet Union: Why not at least explore the possibilities of an internationally managed system for limited strategic defense? Protection from the human and the political effects of an accidental missile launch clearly benefits both the nation whose missile has been launched and the targeted nation.

The general management of an international system, with integrated multinational personnel at all critical operational nodes, could well parallel that of INTELSAT. Guidelines on the application of the system would be established during initial negotiations. Unanticipated needs for new guidelines would be subject to new negotiations. The management entity of the international system must have clear limits to its authority at any given time. The entity, however, must also have clear license to use the technologies transferred to it to destroy or assist in destroying any delivery system whose launch and target were not announced and approved in a timely manner and any delivery system that deviated from an approved flight trajectory.

There is no question that just the technologies available for deployment today can provide a worldwide defense against a few strategic delivery systems. Indeed, the sensors necessary to detect launches are already in use, and ground-based defenses, which are already in place around Moscow, have been demonstrated successfully by the United States. A successful limited system does not require space-based interceptors or any technologies that are at the cutting edge of strategic defense research. Thus, political and bureaucratic resistance to sharing of technology and methods should be minimized.

The United States, Europe, and the Soviet Union must take the lead in the creation of an internationally managed strategic defense system, but nations such as Israel, Japan, China, Korea, Taiwan, India, and Canada clearly have the technological sophistication to participate as additional contributors to the system. Most other nations have the human and political incentives to participate as formal beneficiaries of its protection.

The precedents and experience exist within the world community to establish an international management entity for limited strategic defense without threatening the national prerogatives of member states. A limited international system could in itself be the precedent for developing more extensive defensive capabilities and for extending international cooperation to encompass protection from clandestine or cruise missile delivery of nuclear, chemical, or biological weapons. Learning to work together in new and politically challenging endeavors is but one of the potential benefits of cooperation in international strategic defense.

A renegotiation of the ABM Treaty in light of new technologies and opportunities unforeseen in 1972 would appear to be in the interest of both the United States and the Soviet Union. The creation of an internationally managed system for limited strategic defense should be included in that renegotiation or, potentially, should become the catalyst for renegotiation. Indeed, the additional participation of other nations in this endeavor embodies the full spirit of Article 51 of the United Nations Charter, which is included by reference in the ABM Treaty, and which reads in part as follows:

> Nothing in the present Charter shall impair the inherent right of individual or collective self-defense . . .[15]

A RATIONALE FOR CONTINUING SDI

Up till now, I have concentrated on arguing that internationalizing strategic defense is both ethical and in the national interest. Although I have also given examples of various international technical projects that have succeeded, I have said little about how the practical problem of building support for an international strategic defense system might be overcome. I would suggest that at present the United States should continue carrying out its own research on strategic defense. It may sound ironic, but an unequivocal commitment by the United States to its own strategic defense research would enhance the prospects of international cooperation in this activity. Few nations would wish to be left behind if their own defense and technological development were at issue.

Today, the limitless ocean of space exists as a new frontier and a new challenge for humankind. As the great seafaring nations dominated earlier times, the nations that effectively utilize technology to exploit the economic, environmental, and military advantages related to this new ocean will dominate human activities on Earth well into the next century. Economically and militarily, space is the new high ground as well as the new ocean. Culturally, civilization's imprint on the third millennium, in all probability, will be set by those who meet this challenge.

In contrast to competition on the oceans in the sixteenth century, the competition for domination of space is between diametrically opposed political and economic systems. History, it seems, has already asked the question of whether American and Western political and economic values of freedom and individuality will prevail in this contest, or whether the totalitarian and authoritarian system of the Soviet Union, however dressed up by peristroika, will control the future of humankind. Further, history may be preparing to examine the long-term potential of free enterprise in Asian totalitarian environments, particularly as China attempts this approach, which has been extremely successful in four or five smaller Asian nations. However the current turbulence behind the Iron and Bamboo Curtains and the continuing uncertainty in Asia may be resolved, the values of our competitors will remain fundamentally different from our own. Hence the stakes today are much higher than they were in the sixteenth century: we can hardly expect our values to survive if we fail to move to the frontier of space.

In this competitive environment, the challenge before the United States is to embark into the ocean of space, settle its cosmic islands, develop its economic potential, use it to protect our interests and environment on Earth, and reap, once again, the immense and incalculable benefits of a frontier.

The most fundamental issue before the United States, however, is freedom. Once acquired, nothing becomes more precious or demands greater vigilance. The exercise of freedom requires that all the burdens on the human condition be simultaneously lightened — fear, poverty, hunger, disease, ignorance, and environmental degradation. Space for America really has become a symbol not only of the mission to preserve the freedom we have inherited, but of the mission to spread our inheritance across the Earth. As did our ancestors at the shores of the seas, we now must begin to grasp more firmly the opportunities for freedom and international harmony in the new frontier of space.

As our primary competitor in space, the Soviets have been actively pursuing a comprehensive strategic defense system for many years, a fact too often ignored.[16] Also, the Soviets have used the loopholes in the ABM treaty and U.S. restraint to make considerable progress toward such a system. To think that the Soviets are going to wait for the United States to make a firm decision to create a strategic defense is naive at best and suicidal at worst. The greatest danger in a new round of comprehensive arms control discussions is that we will pause, hoping that the Soviets will pause also.

As the history of the INF Treaty shows, arms control negotiations will remain stagnant until such agreements are clearly in the Soviet leadership's best interest or until the United States makes potentially risky concessions in the interest of progress. The one-dimensional, military–industrial nature of the Soviet economy, an economy otherwise only on a par with the Third World, makes it extremely difficult for the Kremlin to implement any significant reductions in arms production or deployment.

The economic and social disruption implicit in large decreases of weapons-related employment and increases in internal communications and choice probably could not be absorbed without rapidly undermining totalitarian control of the Soviet society. Recent events in the Soviet Republics under peristroika appear to support this thesis. Further, internal bureaucratic and establishment resistance to decreases in military forces will be likely if history is any indication. History strongly suggests that democratic societies absorb the economic disruptions of disarmament much more readily than nondemocratic societies. Thus, a clear and unequivocal commitment by the United States to develop strategic defense systems, coupled with a willingness to share the responsibilities of their use, would make it again in the Soviet's interest to enter into meaningful arms control agreements.

CONCLUSION

The ethical imperative for limited strategic defense has three pillars of support: first, the protection of free institutions and peoples from intimidation or from the tragedy and chaos of mass destruction; second, the use of defensive technology to harness offensive technology moving out of control; and third, the creation of an opportunity for international cooperation in the defense of human beings everywhere. Ultimately, limited strategic defense may grow to global strategic defense in which all nations participate.

One of the simplest and most basic of human concepts underlies strategic defense: self-defense. By ensuring that the world can protect itself from strategic attack, while at the same time ensuring that our conventional forces can defend our interests and the interests of freedom anywhere in the world, a way out of the immoral thicket of Mutual Assured Destruction may be found. The MAD thicket, to paraphrase Dyson, allows the purposeless massacre of innocents, but forbids destruction of weapons in self-defense.

The technical and diplomatic path out of the MAD thicket will be neither easy nor cheap; however, the open fields of mutually assured protection must be reached. Humanity and freedom demand no less.

REFERENCES

1. REAGAN, R. 1983. Transcript of a public statement announcing the Strategic Defense Initiative. New York Times. March 24: A20.
2. MOYNIHAN, D. P. 1979. Reflections on the SALT Process. New Yorker Magazine. November 19.
3. See, for example BEATTY, J. 1989. Reagan's Gift. Atlantic Monthly. February: 62–63.
4. See, for example, KACEWICZ, W. 1987. Affordable, effective SDI is possible by 1994. Wall Street Journal. May 1: 22.
5. ANON. 1988. Nunn redirects antimissile debate. Aviation Week and Space Technology. January 25: 18.
6. WASHINGTON, G. Farewell Address (1796). 1968. Annals of America. Vol. 3: 604. Encyclopaedia Britannica.
7. MADISON, J. Message to Congress (June 1, 1812). 1968. Annals of America. Vol. 4: 314. Encyclopaedia Britannica.
8. MONROE, J. Annual Message to Congress (December 2, 1823). 1968. Annals of America. Vol. 5: 74. Encyclopaedia Britannica.
9. ROOSEVELT, T. Letter to Representative Theodore Burton (February 23, 1904). 1968. Annals of America. Vol. 12: 598. Encyclopaedia Britannica.
10. ROOSEVELT, F. D. & W. S. CHURCHILL. The Atlantic Charter (1941). 1968. Annals of America. Vol. 16: 89–90. Encyclopaedia Britannica.
11. VANDENBERG, A. H. 1948. Vandenberg Resolution. Congressional Record. 80th Cong. 2d sess.: 6053–6054.
12. KENNEDY, J. F. 1963. Television address, July 26, 1963. Congressional Record. 88th Cong. 1st sess.: 13,453–13,455.
13. KENNEDY, J. F. 1961. Policy statement on communications satellites. July 24.
14. ALPER, J. & J. N. PELTON, Eds. 1984. The INTELSAT global satellite system. Progress in Astronautics and Aeronautics. Volume 93.
15. United Nations. Charter (1945). 1968. Annals of America. Vol. 16: 327. Encyclopaedia Britannica.
16. See, for example, Department of Defense. Soviet Military Power: 45–62. U.S. Government Printing Office. Washington, DC.

Reflections on the Moral Debate over the Strategic Defense Initiative

STEVEN LEE

Department of Philosophy
Hobart and William Smith Colleges
Geneva, New York 14456

The moral debate over the Strategic Defense Initiative (SDI) has run its course in a surprisingly short period of time. The debate began less than six years ago when President Reagan initiated the SDI by setting forth a noble moral vision for the role of strategic defenses. The end of the debate has been signaled by the recent news report that the Pentagon is now promoting the SDI for its capacity to destroy Soviet satellites.[1] This concludes the moral debate on an unfortunate note. The history of the SDI is a tale of moral decline. Because the debate, if not the program itself, is at an end, this is an appropriate time to reflect on what it has taught us. There are three lessons to be drawn from the moral debate. The first is the somewhat surprising lesson that morality counts in the area of military policy. The second lesson is less surprising. It is that the fundamental moral problem posed by nuclear weapons remains intractable. The third lesson is that, despite claims to the contrary, there is nothing morally special about the SDI defenses in comparison with other strategic weapons systems.

MORALITY COUNTS

Until the early 1980s, moral concerns seemed to play little role in the public debate over nuclear weapons policy. The flurry of discussions about the moral status of nuclear weapons among theologians in the 1960s did not penetrate very deeply into the public debate. Early in this decade, however, the moral issues raised by nuclear weapons came to be discussed not only in the national media, but at the community level in parish discussions and Nuclear Freeze meetings. One cause for this change was the highly visible debate by the U.S. Catholic bishops on the morality of nuclear weapons, resulting in their pastoral letter, *The Challenge of Peace*.[2] The bishops concluded that nuclear deterrence is morally acceptable only if it is treated as an interim step toward nuclear disarmament.

This public consideration of the moral issues was given further impetus by Reagan's SDI speech. Reagan presented "a vision of the future which offers hope." His plan was "to counter the awesome Soviet missile threat with measures that are defensive," the goal being "to free the world from the threat of nuclear war." He found the need "to rely on the specter of retaliation, on mutual threat" unacceptable. Calling this need "a sad commentary on the human condition," he asked, "Wouldn't it be better to save lives rather than to avenge them?" In doing so, he called upon scientists "to turn their great talents now to the cause of mankind and world peace, to give us the means of rendering these nuclear weapons impotent and obsolete."[3] This speech represented something new in the history of government pronouncements on nuclear weapons policy, a call from the highest levels for a fundamental change in strategic policy argued for in moral terms. No national leader had ever so explicitly and so publicly discussed the inherent moral problem with nuclear deterrence. Reagan not only discussed the problem, but proposed a solution. His noble moral vision was a program for the abandonment of nuclear deterrence.

The SDI has not, of course, been in want of influential critics, but it has, nonetheless, enjoyed in its brief history a substantial measure of political success. Its political success has, fortunately for the program, not been directly proportional to its technological success, but has been due in large measure to Reagan's noble moral vision and the hold it has had on the public's imagination. This is the sense in which the SDI debate has shown that morality counts. The moral debate over the SDI reveals that, in the area of military policy, morality is an important consideration. It is not simply that military policy is, despite the claims of the political realists, subject to moral assessment. Not only is the moral dimension of military policy relevant in assessing that policy, but it is *seen* to be relevant. People treat morality as a relevant and important consideration. The political success of the SDI has shown that public consensus for a military program can be based on moral considerations. The reason that moral concern was not in public evidence before the 1980s is probably that the moral problem of nuclear deterrence seemed so incapable of solution that it could only be discussed in a tone of hopeless resignation. To raise it explicitly would be to admit the inevitable immorality of military policy in the nuclear age. So it was not discussed. Reagan was able to raise the moral problem because he could offer a vision of hope regarding its solution, namely, a policy for the abandonment of nuclear deterrence.

Two comments are in order regarding this argument that morality counts. First, I am not denying that there were other reasons besides the moral ones for the public to be attracted to the SDI. I claim only that the public success of the SDI, following on the undeniably moral justification for the SDI prominent in Reagan's speech, shows that morality has played an important role in the public's support. Second, as I shall discuss below, Reagan's noble moral vision was quickly abandoned by most of the proponents of the SDI. But as the actual strategic and moral rationales for the SDI shifted from the ones proposed by Reagan, his vision has remained at least partly intact in the public's perception, and has been used, sometimes unscrupulously, by the proponents of the SDI as a rhetorical cover to secure a level of public support for the program greater than it would otherwise enjoy. Thus, although Reagan's moral vision has passed from being the actual to merely the apparent rationale for the program, its hold on the public's imagination continues to demonstrate the importance of moral considerations.

THE MORAL PROBLEM IS INTRACTABLE

What precisely was Reagan's moral vision? If we refer back to his words, we can see that it has three, related aspects. 1) The most important aspect of his moral vision was expressed when he referred to "the specter of retaliation," as "a sad commentary on the human condition." Strategic defenses could provide a way beyond this specter by "rendering these nuclear weapons impotent and obsolete." Thus, the first aspect, as discussed above, is his criticism of nuclear deterrence in moral terms and his claim that the SDI could provide a solution to this moral problem by leading to the abandonment of nuclear deterrence. 2) He asked: "Wouldn't it be better to save lives rather than to avenge them?" So, the second aspect of his moral vision is that SDI defenses would save human lives. 3) The SDI would counter the Soviet threat "with measures that are defensive." Thus, the third aspect of his moral vision is that the SDI would replace offensive weapons with defensive weapons.

The second lesson to be drawn from the moral debate over the SDI is that the moral problem of nuclear deterrence remains intractable. This is a consequence of the failure of the first aspect of Reagan's moral vision. What is the moral problem with nuclear deterrence to which he saw the SDI as providing a solution? The policy of nuclear deterrence as practiced by the superpowers is ultimately based on threats to destroy the opponent's society. Put another way, each superpower holds vast numbers of civilians on the other side hostage to the behavior of their leaders. It is morally unacceptable to base a policy on threatening innocent persons, by thus holding them hostage, which is what nuclear deterrence inevitably does. Yet, given the opponent's nuclear threat, there seems to be no other way for each side to provide for its national security. This creates a profound moral dilemma. According to reports, this moral problem concerned Reagan personally. He saw the SDI as a solution to this problem, a way out of the dilemma, because he envisioned the deployment of SDI defenses as a policy for guaranteeing national security that could replace nuclear threats. With the defenses in place, Soviet nuclear warheads could be intercepted before they reached the United States, and deterrent threats of nuclear retaliation would no longer be needed to ensure the safety of the United States from nuclear attack. National security could be preserved while the morally problematic policy of nuclear deterrence was abandoned.

But the failure of this aspect of Reagan's moral vision is by now a familiar story. Nuclear weapons are individually so destructive that the defenses would have to be near perfect in order to neutralize the Soviet threat to destroy American society. Prospects for intercepting nearly all of a large number of Soviet ballistic-missile warheads in the few minutes in which this would have to be done are so daunting as to be virtually nil. Add to this the difficulties created by Soviet efforts to defeat the defenses through the variety of countermeasures that would be available to them, and the task becomes impossible. So, nuclear threats will remain necessary to deter a Soviet nuclear attack. Thus, the moral problem created by nuclear weapons remains intractable. This is not to say that the defenses would not be effective, only that they could not be effective enough to allow the United States to abandon nuclear deterrence.

THE SDI IS NOT MORALLY SPECIAL

The third lesson is that there is nothing morally special about the SDI. To establish this conclusion, we need to consider the second and third aspects of Reagan's moral vision and to follow the moral debate over the SDI after it was recognized that the goal of replacing nuclear deterrence could not be achieved. Once this was recognized, it was seen that whatever defenses were deployed would be part of a policy still based on deterrent threats of nuclear retaliation. As a result, a new strategic rationale for the SDI was propounded. SDI defenses would not be able to replace deterrence, but they could *enhance* deterrence. This rationale was both moral and strategic, for, in enhancing deterrence, the United States would be lessening the likelihood of war. This accords with the second aspect of Reagan's moral vision, for to lessen the risk of war is, in effect, to save lives.

But when Reagan claimed that defenses would save lives, he probably was not thinking of their role in lessening the likelihood of war. Probably he was thinking that the defenses would save lives by intercepting warheads in the event of a Soviet attack. But in the event of war, defenses are unlikely to save many lives, at least relative to the number that would be lost. For example, large American cities, where most of the lives would be lost in a large-scale nuclear attack, probably have a large number of Soviet warheads targeted on each of them. (The United States reportedly has sixty warheads targeted on Moscow.) Nuclear warheads are individually so destructive that most of those killed in an attack that included cities would die as a result of the first one or two warheads that detonated over each large city. So, to save a relatively large number of lives, the defenses would have to be effective enough to guarantee that not even one or two warheads would detonate over each large city, and it is very doubtful that they could be this effective. This is not to say that lives would not be saved, but because the number would not be large in relative terms, lessening the likelihood of war becomes a much more effective way of saving lives.

The moral case for the SDI then rests primarily on the claim that SDI defenses would enhance nuclear deterrence, thereby lessening the likelihood of war. The first thing to note is that the argument for this claim is based on the scenario of a limited Soviet attack against military targets. It is deterrence of this kind of attack that the defenses would enhance. Thus, the argument is based on the assumption that the defenses would primarily provide protection for strategic military targets, such as missile silos. The SDI would enhance deterrence not by its capacity to protect American lives in the event of an attack, but by its capacity to protect missiles. This is in sharp contrast to Reagan's moral vision, which foresaw the defenses being deployed to protect people, not missiles. In any case, if protecting missiles would enhance deterrence, then the defenses would *indirectly* protect people by making war less likely. What is the argument that the defenses would enhance deterrence?

Proponents of the SDI argue that the defenses would enhance deterrence by creating greater uncertainty about the results of a limited Soviet attack on military targets and by requiring that the Soviets commit more of their forces to such an attack.[4] If the Soviets are less certain about the outcome of an attack or have to expend more of their forces to undertake it, they are less likely to undertake it, and so will be better deterred. More generally, the argument is that, in intercepting some of the warheads in a limited Soviet counterforce attack, the defenses would help to ensure that the United States could more effectively retaliate, and hence would better dissuade the Soviets from initiating such an attack. I do not now want to address the soundness of this argument.[5] What I want instead to show is that this argument is basically the same as the argument advanced for counterforce modernization programs in general.

The rationale for counterforce modernization programs, such as the MX land-based missile or the Trident II submarine missile, is also the enhancement of deterrence. It is argued that these weapons systems will provide the capacity for a more effective retaliation against a limited Soviet attack, thereby making the threat of retaliation more credible, thus enhancing deterrence. But this is, in general terms, also the argument for SDI defenses. Of course, the SDI and the MX would achieve this end in different ways. Defenses threaten to intercept warheads in flight, while offensive weapons threaten to destroy warheads on the ground as part of a counterforce retaliation (or a preemptive attack). But in each case it is the contribution that these weapons would make to the overall capacity for more effective retaliation that is said to enhance deterrence.[a] The moral argument that deterrence will be enhanced and the likelihood of war lessened is the same in each case. Thus, the SDI is not morally special. Once it is recognized that defenses could not replace nuclear threats, the moral grounds in favor of the SDI re-

---

[a] Counterforce ballistic-missile warheads have been referred to as "pre-boost phase" defenses.

duce to those in favor of counterforce weapons systems in general. The basic point is that the SDI is, as far as moral considerations go, simply another counterforce weapons system.

But perhaps this claim is too strong. Even though SDI defenses are a counterforce weapons system, still, it seems, the way in which they operate would mark them as morally special. This raises the third aspect of Reagan's moral vision. The SDI is defensive rather than offensive, and this implies, Reagan seemed to be arguing, an important moral difference. Of course, counterforce weapons systems normally classified as offensive can be used defensively. For example, MX missile warheads might be used in a defensive war for damage limitation. But because offensive weapons systems can also be used for aggression, for example, in an unjustified first strike, they are not, let us say, *purely defensive*. Let us stipulate that a weapons system is purely defensive only if it cannot be used for an aggressive purpose. A weapons system is defensive if it has defensive uses or is in fact generally used defensively, but it is purely defensive only if all of its potential uses are defensive. It seems that Reagan envisioned weapons that would be purely defensive in this sense. If SDI weapons would be purely defensive, they would indeed be morally special.

Weapons that are purely defensive in our sense are morally special because they cannot be used for aggression. They are reactive and cannot be used to initiate military action. But also of moral importance is a related difference peculiar to SDI defenses. This difference may be part of the second aspect of Reagan's moral vision. When he spoke of the defenses as saving lives, he may have been thinking not only of American lives, but also of Russian lives. For SDI defenses in operation would kill no one. Because missile warheads are pilotless, the defenses would not even kill Soviet military personnel. This difference also seems to mark the SDI as morally special, for other counterforce weapons systems, even when used defensively, would kill many people, including civilians, as "collateral damage." Thus, it seems, there are two features that may make the SDI morally special: SDI defenses would be purely defensive, and they would kill no one.

We must now ask whether SDI defenses would in fact have these two features and, if so, whether having them implies that the SDI is morally special. First, consider the claim that SDI defenses would be purely defensive. Here the recent news report that the SDI is being promoted as an antisatellite weapon becomes relevant. It is not surprising that the SDI would eventually be touted for its antisatellite capability, there being serious doubt whether SDI defenses could be effective enough to do much in way of enhancing deterrence, let alone allowing the abandonment of nuclear deterrence. Although intercepting warheads is very difficult, knocking out satellites is comparatively easy, so the antisatellite mission is one that the defenses would surely be able to achieve.

But, in repetition of the earlier pattern, a new strategic rationale leads to a more problematic moral rationale. When the strategic rationale for the

SDI shifted from making nuclear weapons obsolete to enhancing deterrence, the moral case for the SDI suffered. Now, with the strategic rationale shifting from enhancing deterrence (through intercepting Soviet first-strike warheads) to providing an antisatellite capability, the moral case grows weaker still. For their having an antisatellite capability shows that SDI defenses would not be purely defensive. The destruction of satellites can be an act of aggression. In fact, a very effective way for the United States to initiate a surprise nuclear attack would be to destroy Soviet reconnaissance satellites. So SDI defenses would not be exclusively reactive. Their use in war would not have to wait upon the launch of Soviet missiles. Because SDI defenses would have such an aggressive use, they would not be purely defensive.

There is, however, a larger issue involved in the claim that SDI defenses would be purely defensive. One cannot determine whether a weapons system is purely defensive by considering that system only by itself, without reference to the other weapons with which it would be used. For the moral character of a weapon is determined by how it would be used in practice. If, in practice, the weapon would be used with other weapons, its moral character—for example, whether or not it should be considered purely defensive—cannot be determined by how it might be used if it were used alone. In war, SDI defenses would be used along with other counterforce weapons systems. If the defenses could be used along with these other weapons as part of an act of aggression, then the defenses cannot claim the moral advantage of being purely defensive.[6]

Consider the analogy of a bulletproof vest. If the wearer of a bulletproof vest is unarmed, then the vest would be purely defensive, in that the vest could not then be used as part of an act of aggression. But an act of aggression with a gun can be facilitated by the wearing of a bulletproof vest, so a vest on someone who is armed, though it might still be used defensively, would not be purely defensive. Even apart from their antisatellite capability, SDI defenses, when considered along with the other counterforce weapons systems with which they would be deployed, would have an aggressive use. SDI defenses that were moderately effective would enhance the capability of the United States to threaten or to carry out a successful first strike. For the defenses could provide for U.S. cities substantial protection from the "ragged retaliation" that could be managed by the Soviets following a U.S. counterforce first strike.

The need to consider the context in which a weapons system might be used shows that the second feature of SDI defenses, the fact that they would by themselves kill no one, does not make the SDI morally special. The morally relevant question is whether people would be killed when the defenses are used, not whether the defenses themselves would be directly responsible for any of the deaths. If a sniper succeeds in killing more people than he or she otherwise would have killed because of protection from a bulletproof vest, the wearing of the vest is not morally commendable, even though the vest itself killed no one. SDI defenses might kill no one, but this does not make them morally special in comparison with other counterforce weapons systems. SDI defenses, as well as other counterforce weapons systems, could be used either aggressively or defensively. So the argument that the SDI is morally special fails.

The moral debate over the SDI as a peculiar and unique military program has come to an end because it has now become clear that the moral rationale for the SDI is no different than the moral rationale for counterforce weapons systems in general. In other words, the moral debate over the SDI is now seen to merge with the larger moral debate over the deployment of counterforce weapons. In general, what could be said in moral terms in favor of or in criticism of the SDI could be said as well about other counterforce weapons systems. This has been revealed by the news report about the SDI being promoted for its antisatellite capability, for this report has made clear what should have been clear earlier, that the SDI could be used aggressively as well as defensively. The soundness of the moral argument for counterforce weapons systems in general is another matter, and one that I cannot discuss in this paper. I have only sought to show that there is nothing morally privileged about the SDI. The basic lesson of the moral debate over the SDI is that after six years, everything looks the same.

REFERENCES

1. Broad, W. 1988. U.S. Promoting Offensive Role for "Star Wars." N.Y. Times. November 27: A1.
2. National Conference of Catholic Bishops. 1983. The Challenge of Peace: God's Promise and Our Response. U.S. Catholic Conference. Washington, DC.
3. Reagan's speech is reprinted in many places. For example, see Lackey, D., Ed. 1989. Ethics and Strategic Defense: 36-38. Wadsworth. Belmont, CA.
4. For one source of this argument, see Sloss, L. 1989. The case for deploying strategic defenses. *In* Nuclear Deterrence and Moral Restraint. H. Shue, Ed. Cambridge University Press.
5. Lee, S. 1988. Morality, the SDI, and limited nuclear war. Philos. Public Affairs **16**(1): 15-43.
6. For a discussion of the moral argument for the SDI based on the claim that it is purely defensive, see Kavka, G. 1987. Moral Paradoxes of Nuclear Deterrence: 147-164. Cambridge University Press.

Strategic Defense Initiative Research
A Gray Area for Ethics

AANT ELZINGA

Department of Theory of Science
University of Gothenburg
S-412 98 Gothenburg, Sweden

INTRODUCTION

The relationship between ethics and research in the military is much more complex than it was almost two decades ago, when the Vietnam War prompted a strong movement criticizing the military connection. Military technology did not base itself, to the extent it does today, on the most advanced research available.[1,2] One reason for the difference today is the nature of the new technologies that are employed, as for example in the Strategic Defense Initiative (SDI) program. Whereas research for the military previously often had a more applied character, today the character of the new and emerging technologies involves much more basic research. High technology, as it is often called, comprises a number of technological clusters in the areas of microelectronics, new industrial materials, and biotechnology. In these areas, there is a much stronger symbiotic relationship between basic research and technological development. Consequently, with science (and not just military science) becoming more mission oriented, channeling resources to promising fields of basic research assumes greater importance. For example, "foresight" exercises are carried out by review panels that are much more obviously consensus oriented and use more openly subjective methodologies than any of their predecessors in technological forecasting. Foresight for science and technology is carried out in industry, in governmental agencies, and in some nongovernmental agencies. The objective is to scan scientific and technological developments, to find strong and weak points, and to identify opportunities and threats (including market threats attributable, say, to a competitor's advantage in scientific potential or in applying cutting-edge technology).

The symbiotic nature of the science–technology relationship is also evident in the introduction of a new category to add to the traditional trichotomy of basic research, applied science, and developmental work. The new category is called strategic basic research, or just strategic research. John Irvine and Ben Martin define it as "basic research carried out with the expectation that it will produce a broad base of knowledge likely to form the background to the solution of recognized current or future practical problems."[3] Strategic research also appears as a new category within research councils and mission-funding agencies for accounting and steering purposes. In research councils, it is sometimes referred to as targeted research.

Targeted areas in many countries are connected with microelectronics, biotechnology, and materials science, but also with women's studies, environmental protection, and research on aging. The significant point is that fields of research are being defined, not primarily from the internalist point of view of disciplinary advances, but rather from the point of view of anticipated utility in achieving externally defined goals like developing a particular technological system or clusters of new technologies with potential for economic growth, military-political objectives, or social welfare. Because the concepts used for accounting, together with foresight exercises under the auspices of military programs, can thus steer the agenda of basic research, a gray area confronts those who would like to avoid doing research for certain military ends, such as those of the SDI program. How can one know if the research one is doing will or will not dovetail into military technology? This complicates the ethical problem, both at the individual level and the institutional level. In the case of SDI, a boycott pledge movement has gathered about 6000 names, half of them faculty in areas of physics, mathematics, computer science, and so on, in order to take a stand of refusing to lend credibility to a megaproject that has been deemed unsound for various moral as well as technical and scientific reasons. This indicates to me that the ethical question must also be posed in science policy terms. What is the orientation toward military ends doing to the conditions of basic science and, ultimately, the competitiveness of industry in an international arena?

MORE SPIN-IN THAN SPIN-OFF

The critique of the military connection during the Vietnam War concerned financial and institutional control, as well as moral and ethical questions. Today these issues are still very much part of the debate, but in addition, and equally important, are questions about more indirect influences—which, for example, may skew research agendas, as when powerful mission-oriented research and development programs in the military sector exert a kind of magnet effect on institutions carrying out basic research.

Even those who are otherwise comfortable receiving Department of Defense funding are beginning to have doubts:

> If most of the funding for developing the technological base in the U.S. is controlled by a small number of government agencies with specific mission objectives [the Department of Defense, the Department of Energy, and the National Aeronautics and Space Administration], does this preclude important scientific and technological advancements that serve other areas of civilian and social need?[4]

It has been argued that the skewing of basic research reduces the competitiveness of industry in the long term, unless of course spin-off is a significant factor. As already indicated, it also makes the question of ethical aspects more difficult to handle. The arguments of fifteen years ago had to do with visible indicators of militarization of research. Today, because of the research-intensive character of many emerging technologies, there is a growing "gray area of research," to use the term of the Corson report.[5] In this gray area, it is impossible to distinguish the military from the nonmilitary. The Corson report is vague on this point, speaking of research involving technologies that meet a number of criteria. Among these criteria are rapid growth and short lead time between basic research and application. The Corson report also refers to technologies that have dual uses and involve process- or product-related techniques, where moreover the transfer of such technologies would give the Soviet Union a significant near-term military advantage. The definition gives a lot of leeway for differences of interpretation in specific cases because in fact it is sometimes impossible to see where a civilian realm ends and where a military or even more broadly defined realm of national security interests begins. Potentially the payoff of basic research can lie in either.

In this connection, the term spin-off is also inadequate, because it assumes a clear-cut movement from the military to the civilian realm. With the strategic research having dual uses and with long-term mission orientation, we have another phenomenon, going in the opposite direction. Some analysts have preferred to call this "spin-in," that is, movement into the military from the civilian realm (FIG. 1).[6] It is this possibility of a spin-in effect of basic research done under purely civilian auspices that magnifies the problem of the ethical aspects. Researchers could easily work on civilian projects while being unaware that these projects could have important military applications and that these apparently civilian projects could, furthermore, be integrated with military projects by high-level research managers having the resources (for example, computerized facilities) to coordinate a broad range of research activities. It also seems to be the case, as argued by the Council of Economic Priorities,[7] that the benefits of spin-off are much exaggerated. The Council sees a dwindling technological leadership. This may be even more serious as spin-in into the military sector continues, as economic and personnel resources are spent, and as scientists become cynical or demoralized. As I shall argue, military mission orientation of the type SDI research represents has a lot of negative effects on the climate of basic science. In Sweden, we have found similar types of effects in other areas of (nonmilitary) mission funding; in the military they become aggravated.

SDI: A CONTROVERSY WITH MANY DIMENSIONS

The introduction of the strategic defense concept and the actual research programs have precipitated a wide-ranging controversy. In fact, there is a whole bundle of interwoven controversies surrounding SDI. Opposition manifests itself both in debate and in concrete action, the latter in the form of a

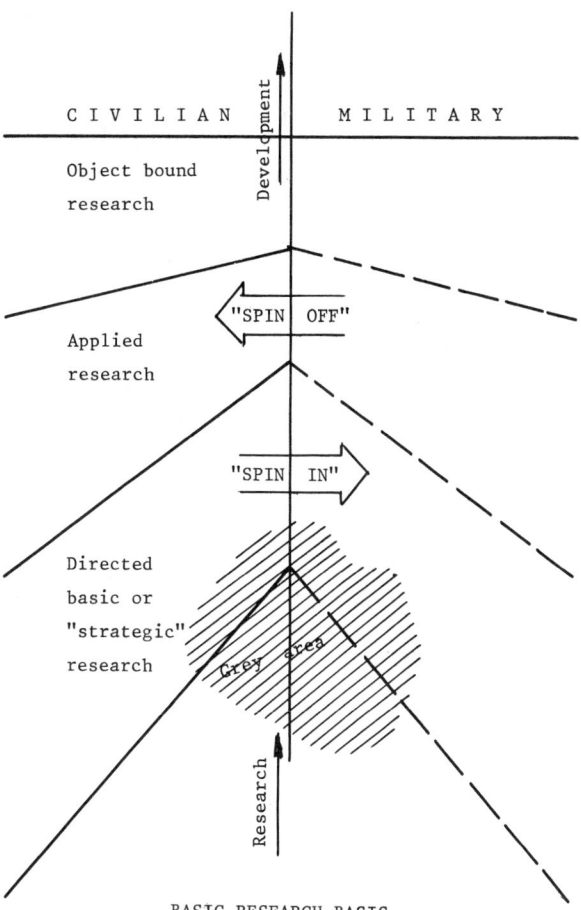

FIGURE 1. A diagram showing the range of activities from research to development, with particular emphasis on the directions of spin-off and spin-in with respect to the civilian and military sectors. Adapted from Hagelin.[1]

boycott pledge movement that has even spread to Britain. The acceptance of SDI funds is seen as a political act by quite a number of academics today.[8]

Looking at the various issues within the controversy, one can identify several dimensions. These include military–strategic, military–political, and security arguments; scientific and technical arguments; ethical and moral arguments; as well as arguments relating to economic and industrial policy.[a]

---

[a] For giving me insight into the complexity of the controversy, I am indebted to Wilhelm Agrell and Kent Lindkvist at the Research Policy Institute, University of Lund, Sweden. Lindkvist's chapter in an upcoming book in Sweden has been particularly helpful, together with documentation provided by Gunnar Eriksson and Johan Swahn at the Technical Peace Research Unit, Technical University of Chalmers, Gothenburg, Sweden.

Briefly, the military strategic controversy revolves around the issue as to whether SDI is really a defensive system, or if it is potentially a more sophisticated form of offense, incorporating the doctrine of Mutually Assured Destruction (MAD). Rivalry for seeking military superiority determines the nature of the U.S.–Soviet arms race in space, and under circumstances of acute confrontation scientific and technological advances certainly open up new fields in the arms race. Given time and technology and the eventual testing and deployment of space weapons, strategic offensive weapons systems as well as defensive ones could thus hardly be prevented. This is a serious destabilizing factor for peace.

The military–political and security aspects raise the question of whether SDI should be considered research, or whether it should be considered weapons production. The position taken on this has a bearing on whether one sees SDI as contravening the Antiballistic Missile Treaty of 1972 and possibly also conflicting with earlier United Nations resolutions against bombs in space (1963) and nuclear weapons in space (1967). Also related are different perceptions of the relative technological prowess of the Soviet Union and its mobilization of countermeasures, which would prompt further countermeasures by the United States, thus escalating the nuclear arms race between the two superpowers and bringing it into space. This would open a new era of tension and increase the danger of starting a global war.

Crucial in the foregoing controversy is the issue of the survivability of a fully deployed SDI system. This gets into the heart of a complex scientific and technical controversy, which also touches the question of feasibility and the time perspective for a realistic decision on possible deployment. A report issued by the American Physical Society on directed-energy weapons, according to many researchers, put a clincher on this controversy by stating that at least ten more years of research is needed before it makes sense to talk about a decision concerning the ultimate feasibility of directed-energy weapons in space.[9] Another sore point is battle management, where critics either argue that adequate software programs could never be realized, or point out that any software programs that could be realized could never be tested under realistic conditions.

Ethical and moral dimensions of controversy concern the character of the SDI concept itself, whether it really is a progressive step forward for mankind because it would make nuclear weapons obsolete, or whether in fact it is a further step in the direction of barbarism. There is also the ethics of pursuing basic research that is related to an end result that might be judged scientifically and technically unsound, that is, as being bad science. Further, there is the added complication that the military connection brings with it potential censorship, which may be understood as a basic infringement on the traditional academic values of free inquiry and the free flow of information.

Regarding the economic and industrial policy dimension, there are both those who see SDI as a boon, contributing to the cutting edge of modern technological competence and thus enhancing the prospects of market competition in the 1990s, and those who maintain that SDI constitutes a drain on resources needed elsewhere, undermining the very foundations of science and technology needed for competitive industrial strength in the future.

The various controversies are embedded one in the other, overlaid or underpinned by important ideological, political, and professional stakeholder interests that pull in different directions. For the theory of science or sociology of science, the controversy clearly reveals the texture of interrelationships between the social, political, and cognitive dimensions of science in society, making it a particularly challenging site for "controversy studies," a subfield of "science studies."[10]

What I am concerned with here is only one aspect of the SDI controversy: the effect the program has on basic research in the universities.

SDI AND THE MILITARY SECTOR

Department of Defense funding has grown nearly twice as fast as any other federal source of funds for university research and development since 1980. SDI is accentuating this trend. SDI funding is projected to grow to $6.2 billion by fiscal year 1988, but will probably fall considerably short of this. The spending for the current fiscal year (1987) has been put at $3.5 billion.[11] This is part of an anticipated expenditure of $30 billion over the years 1985–1993. John Pike of the Federation of American Scientists has estimated that this strategic defense funding will be responsible for an additional funding of over $11 billion (or $10.3 billion, adjusting for inflation) in defense spending over the years 1985–1989. In any case, after three years, the SDI budget has nearly tripled; the total allotment to basic research during the current year is probably in the neighborhood of $135 million (counting a portion allotted through the Department of Energy). By all counts, it is a crash program that should have an effect on the conditions of basic research in the country. The new and emerging technologies that have been targeted are well known. They include laser physics and chemistry, computer science, mathematics, microelectronic engineering, new materials research, and a few other fields. Relatively speaking, SDI will use less than half the research and development resources required for the Apollo project, which in its peak year meant an employment of 92,000 scientists and engineers by the National Aeronautics and Space Administration. A difference perhaps is that SDI involves a proportionally greater basic research component. Zegveld and Enzing reckon that SDI could employ about 12,000 scientists and engineers during 1987, which is estimated to be 5% of the total science and engineering workforce in the nation.[12]

In the United States, currently about 5% of the SDI budget accounts for mission-oriented basic research in universities and small business firms.[13]

The SDI program is comparable to major crash programs within the science policy sector. As such, it may be reckoned to have both an industrial policy function and a military objective. SDI is sometimes seen in Europe as a covert industrial policy program. Indeed, shortly after its announcement, Mitterand in France launched his proposal for Eureka, a civilian program for industrial research and high-technology development in Europe that in part covers the same areas of technology and science investments found in SDI. The motivation for the French initiative was SDI itself, which was understood as a threat, creating the potential for a widening technological gap between Europe and the United States.

In sum, the financial development of the SDI programs has been rapid by any count. It was noted that expenditures would be about $30 billion over the years 1985-1993, when a decision for deployment must be in hand. Deployment cost estimates vary anywhere from around a few hundred billion dollars to $800 billion or more.

EFFECTS OF THE SDI PROGRAM ON BASIC RESEARCH

Overall Imbalance

The notion that a large military budget is harmful to the economy as a whole is not new. Mary Kaldor of the University of Sussex argues that the concentration on research and development in the military sector leads to a technical dynamic where more and more effort is expended for diminishing returns in military effectiveness.[14] That technological overdevelopment has a skewing effect on the overall balance of scientific and technological development is noted by John Tirman.[15] He estimates that in 1980 about 30% of all funds for research in universities, colleges, and federally funded research centers came from the Department of Defense, the Department of Energy's military programs, and the National Aeronautics and Space Administration. This is seen to constitute an imbalance—and it is aggravated by SDI. Clark Thomborson from the University of Minnesota argues that the growth of the Defense Department's role in computer science not only distorts the direction of the field, but also lessens the potential for spin-offs that could serve valuable commercial and other social uses.[16,17] Moreover, it inhibits advances in scientific foundations, and thus undermines rather than promotes social and industrial potential. Richard Ennals documents a similar point regarding British involvement in SDI research and development.[18]

Uneven Development

The uneven development of fields of science has been noted by Tirman, who writes that major universities, many of which enjoy high levels of DOD support, may design curricula to satisfy particular needs of defense research. The funding mechanism both for individual researchers and institutions encourages a pursuit of science and engineering that contains defense applications, thus skewing, however subtly, the intent and conduct of a variety of research and development projects.[19]

A preliminary study of funding patterns in computer science notes that

> there is also concern that certain types of research [such as theoretical computer science] are being shortchanged by the emphasis on more applied areas. It has also been suggested that some people who should be funded are not receiving adequate support. Evidence from our initial interviews and other sources lends some support to these concerns.[20]

In computer science it is estimated that 60% of research funding comes from the Department of Defense, and that at some of the top schools, like Stanford, MIT, and Carnegie-Mellon, military funding levels may be as high as 75–80%.[21] A question arising from the Swedish experience with the policies of different sectors is to what extent certain research groups or departments may become more strongly bound, socially and cognitively, to bureaucratic agencies than to intrascientific communities.

A long-term consequence of SDI funding has been taken up by Congressman George Brown, who warns that if deployed the system will constitute a brain drain on space research and development and on civilian space technological capability.[22]

Less Emphasis on Basic Research

A tendency of research to be skewed toward applied problems has been noted by various critics of defense funding. MIT Provost John M. Deutsch, a member of the Defense Research Board, has been critical of the Pentagon, claiming it has demonstrated a bureaucratic short-sightedness and a

> narrowness in its choice of projects, supporting only those with apparent military promise, rather than first and foremost selecting those that are expected to yield new understanding and insight.

The Office of Innovative Science and Technology (IST), which promotes SDI, is also seen as reflecting the heavy-handed military bureaucratic approach, "redirecting research on campus to military purposes."[23]

It is clear from the IST Office's initial briefing paper that its express intention is to create a pull toward applied science that fits military strategic plans.[24] The paper pointed to seventeen narrow areas of research, which are targeted and described, and stated that the purpose of the IST Office is to

> mount a mission oriented, basic research program that drives the cutting edge of the nation's science and engineering effort in a direction that supports existing SDI technological development thrusts and points the way for future initiatives.

A prominent SDI critic, Vera Kistiakowsky of the Physics Department at MIT, comments:

> A consequence of this structure is that individual research grants will only be made for brief introductory periods and then will be expected to fit into or develop into a consortium or program. Unlike true basic research, the funding decisions will not be made on the basis of scientific merit alone, the overriding criterion being usefulness to the program.[24]

The inversion of internal peer review and external mission control criteria perceived here is equivalent to what we have termed epistemic drift. It is a phenomenon not restricted to the IST Office; we also find it explicitly stated as a goal by people in the Defense Department's Defense Advanced Research Projects Agency (DARPA). The director of DARPA, in an annual report for the strategic computing program, described the purpose of his program as follows:

> Since the start of this program in November 1983, DARPA has pursued a research and development strategy that is designed to lead the nation's entry into the vast, promising domain of machine intelligence and advanced computing technologies. A key element of this strategy is to foster application-oriented research that contributes to technological advantages which have historically sustained our defense posture in the face of a growing numeric imbalance of forces. These military applications collectively pull the basic research efforts and focus them in the most rewarding areas.[25]

It is evident that DARPA also has industrial policy ambitions. However, unless a university department has solid basic faculty funding for fundamental problems, the kind of pull aimed for by DARPA would tend to erode the receiving institution's basic research profile in the field concerned. This is our experience with various government programs in Sweden.

Slackening of Quality Control

J. Gollon warns that if the portion of federal funding that is peer-reviewed by the National Science Foundation declines even more, while military funding continues to dominate, we may witness the deterioration of the

> open, usually higher quality, basic research that has brought the U.S. preeminence.
> ... Will the price for the short-term Star Wars-induced prosperity of U.S. scientific institutions be their long-term decline in quality?[26]

This kind of concern is aggravated by two factors. One is the attitude of SDI officials; the other is the additional interference with the peer-reviewing process that is introduced by military demands for secrecy or limiting access to information.

SDI officials like James Ionson, the director of the IST Office, are keen to buy political legitimacy. For example, Ionson has stated that it is probably something that has never been done before. But the office is trying to sell something to Congress. If we say that this fellow at MIT will get money to do such and such research, it's something real to sell.[27]

David Parnas, a software expert who turned his back on SDI, has observed that the Department of Defense has some excellent people, but that they are not in the funding agencies. He describes the IST Office as

> a typical organization of technocrats ... so involved in advocacy to the program that it cannot judge the quality of research involved.[28]

The other point about the slackening of quality control has been stated generally by John Tirman, who points out that

> the normal channels of communication, criticism and refinement of data and methodology are closed to the military scientist. Moreover the secrecy enforced within a specific project—the limit on the "need to know" the purpose of the research or even the next step of application—distorts the natural environment and scope of the scientist. The narrowness thus engendered can be a crippling constriction of a scientist's development and maintenance of skills.[29]

The point is that interference with the free flow of scientific discourse in the long run becomes counterproductive, as does state control of science, be it exerted in the creation of mission-oriented programs or in other forms. The step to a form of Lysenkoism is not far. Perhaps this is what Charles Schwarz, a physicist in Berkeley,[b] had in mind when he characterized fellow physicists supporting SDI as

> the American counterparts of Lysenko—political ideologues and hucksters in science acting under the patronage of a powerful chief of state.[30]

Sensitivity to Changes in the Political Arena

The Department of Defense took seven months to complete a prepublication review of a report prepared by the American Physical Society on laser and particle beam weapons.[31] Although the report was eventually cleared, during its review SDI advocates were apparently gearing up for a shift of tactics, because shortly after the appearance of the report the Defense Acquisition Board approved an SDI plan to speed up the testing of six key technologies more mature than those in the laser field. Even though the SDI lobby has not given up the more utopian visions criticized in the report, in practice there is an admission to partial defeat. Now the phased deployment program, which was already a revision of an earlier timetable and plan, has been redirected to early deployment. The tactic is obviously intended to cement SDI

---

[b] Schwarz, who apparently considered the consequences of his own position as professor and teacher of a new generation of physicists under the present circumstances, eventually quit his position at Berkeley.

so that the next White House administration will be saddled with it, obliged to follow the present course.

For research, this means a sudden reorientation in priorities. This is evident from the dramatic shift from the directed-energy weapons to the kinetic-energy weapons budget. The development of the more mature technology of kinetic-energy weapons comes into the foreground, and one finds a reemphasis on having a layered defense system, more reminiscent of the "high-frontier" concept of General Daniel O. Graham. The push for an early deployment date hinges on the development of a system of large garage-like containers with many kinetic-energy cannons, the so-called space-based kinetic kill vehicles (SBKKV), that SDI advocates hope to get ready for deployment in 1994 or 1995. SDI funds for the more long-term innovative technologies are accordingly cut back as political tactics change.[c] Critics have remarked how the latest move represents a reorientation that follows the twists and turns of the SDI program at a time when it was beginning to sputter, a victim of its own erratic course, serious technical uncertainties, and vigorous political opposition.[32,33]

The reorientation means that physicists will play a reduced role, both as innovators and probably as critics too, at least for the time being.

Opportunism and Seeds of Conflict

While head of research and development at the Department of Defense, Under Secretary Donald A. Hicks, in his testimony before an Armed Services Hearing, stated that he was

> not particularly interested in seeing department money going someplace where an individual is outspoken in his rejection of department aims, even for basic research.[34]

The Defense Department, by demanding prepublication reviews and security clearances, may also intimidate some scientists and prompt them to exercise self-censorship, or to withhold papers from public international scientific forums. The previously cited study on the situation in computer science points out how cynicism and conflict may be generated:

---

[c] Richard Ennals, in a talk in Stockholm on February 1, 1988, pointed out how different stakeholder interests, in promoting different concepts for the SDI program, try to influence its direction. Reagan is the only one who still has the idea of a global shield. U.S. military interests prefer focusing on a limited shield to protect U.S. missile silos. Industry, its appetite whetted, hopes for large contracts for placing SBKKV platforms in the sky. A more moderate line, more attractive to wider interests, is to concentrate on ground-based laser weapons stations. This is also of interest to some European governments, and it is what the Soviet Union has admitted to having. The British, for example, believe this would give some defense for their own country. The limited shield concept only serves the United States.

DoD funding has an effect also on those not so funded. Since some researchers who feel uncomfortable receiving DoD grants have trouble finding alternative sources, an indeterminate number of otherwise competent or even superior CS professionals may be limited in their work, or even end up leaving the field itself. Some examples of this have already come to our attention, though the data is still anecdotal. . . . Related to this is the issue of academic freedom and the creation of an unhealthy environment of mistrust and cynicism within the academic CS research community. For example, there are indications that some researchers misrepresent the nature of their work in order to obtain funding from mission-oriented sources.[35]

HYBRIDIZATION AND EPISTEMIC DRIFT

One of the organizational features of work in the SDI program is that consortiums are established in which the activities of university and industrial researchers are coordinated and directed toward solving particular problems—developing, for example, parts of a technological system or systems. It is interesting to note that the Pentagon, apparently in hopes of getting around a congressional ban on establishing a think tank devoted to SDI, is setting up a consortium of federally funded research and development centers that could perform the functions of an SDI think tank.

The creation of consortium networks of this type is a specific manifestation of what we have called the generation of hybrid communities, which have accountability systems and reputational repertoires that differ from those of academic science. It is evident that such structures, linked to mission objectives having externally defined scientific and technological goals, can come into conflict with the social and cognitive structures of the basic research communities. Thus a researcher's involvement in a mission-oriented program poses a potential conflict of interests, with the researcher being caught between the academic research community and the hybrid community. The research agenda is oriented toward and may be subordinated to the logic of the latter—which, as I have argued, would constitute an inversion of internal and external criteria of assessment in the research process. This epistemic drift also brings with it a concomitant shift in the meaning of science.

There is, in a broad sense, a similar problem even with the new Engineering Science Centers of the National Science Foundation. The National Academy of Sciences, in a panel on science and technology centers, recommended the organization of such centers on university campuses, so as to benefit the basic research infrastructure. The panel, however, warned that if such centers committed themselves to a "short-term focus on commercial application," they could find themselves caught in a "self-defeating drift."[36] It is clear that this is in the interest neither of academic science nor of the large corporations who want a "window" on academic research.

SUMMARY

The arguments concerning the militarization of research fifteen years ago in connection with the war in Vietnam had to do with more manifest indicators of a military link. Then it was a question often of applied research, and the physical or institutional control over such research. Today we are dealing with new technologies and their basic research underpinnings. Because of the character of these new military technologies, we are facing extensive gray areas. It is no longer possible to so clearly discern boundary lines. Thus it becomes more difficult also to clearly determine when one should hold off from a certain type of research for ethical reasons. Such boundaries also blur.

As research and development has become increasingly target or mission oriented, externalist assessments relating to social and political problem definitions have taken on greater prominence in defining what is important. An epistemic drift has been noted, a shift whereby assessments of quality or utility by outsiders and nonscientists have become more common. A corollary to this epistemic drift is the erosion of the meaning of basic research in the direction of practice-relevance, a problem that concerns laser physicists as well as social scientists in newly emerging disciplines.[37]

At a very general level, we might speak of a change in the social paradigm of science, a change that has become more obvious in the last couple of decades as the role of the state has expanded.

Finally it is clear that the concept of epistemic drift has to be divested of ideological overtones, particularly those that would equate it with a purist assumption of autonomy and disembodied science. Science as a whole is founded as a social mandate, and autonomy is always relative. What is at issue here is rather the integrity and meaning of scientific enterprise.

REFERENCES

1. HAGELIN, B. 1983. Militär forskning, vad, var, hur? *In* Två uppsatser. Research Policy Institute, University of Lund.
2. AGRELL, W. 1983. Växer innovationerna ur gevärspiporna? *In* Två uppsatser. Research Policy Institute, University of Lund.
3. IRVINE, J. & B. MARTIN. 1984. Foresight in Science: Picking the Winners. Pinter. London.
4. YUDKEN, J. S. & B. SIMONS. 1987. Federal funding of computer science: A preliminary report. CPSR Newsl. 5(3): 2.
5. National Academy of Sciences. 1982. Report of the Panel on Scientific Communication and National Security (D. Corson, Chairman). National Academy of Sciences Press. Washington, DC.
6. HAGELIN, B. 1983. Två uppsatser: 18.
7. Council of Economic Priorities. 1988. Star Wars: The Economic Fallout. Ballinger. Cambridge, MA.
8. BROAD, W. J. 1985. Star Warriors. Simon & Schuster. New York, NY.

9. American Physical Society. 1987. Science and Technology of Directed Weapons. American Institute of Physics. New York, NY.
10. ENGLEHARDT, H. T., JR. & L. CAPLAN, Eds. 1987. Scientific Controversies: Case Studies in the Resolution of Disputes in Science and Technology. Cambridge University Press.
11. ANON. 1987. Aviat. Week Space Technol. April 13: 118.
12. ZEGVELD, W. & C. ENZING, Eds. 1987. SDI and Industrial Technology Policy. Pinter. London.
13. For an overview in Swedish, see Sveriges Tekniska Attacheer. 1986. SDI-ett strategiskt försvarsinitiativ. Stockholm.
14. KALDOR, M. 1981. The Baroque Arsenal. Andre Deutsch. New York, NY.
15. TIRMAN, J. 1984. The Militarization of High Technology. Cambridge, MA.
16. YUDKEN & SIMONS. 1987. CPSR Newsl. 5(3): 7.
17. THOMBORSON (formerly THOMPSON), C. 1986. Military direction of academic CS research. Commun. ACM 29(7): 583-585.
18. ENNALS, R. 1986. Star Wars: A Question of Initiative: 86. John Wiley & Sons. New York, NY.
19. TIRMAN, J. 1984. Militarization of High Technology: 216.
20. YUDKEN & SIMONS. 1987. CPSR Newsl. 5(3): 6.
21. YUDKEN & SIMONS. 1987. CPSR Newsl. 5(3): 12.
22. ANON. 1987. Aviat. Week Space Technol. March 9: 118.
23. GOODWIN, I. 1986. Deutsh suggests ONR as a model for a DoD research agency. Phys. Today. December: 50-51.
24. KISTIAKOWSKI, V. 1986. SDI and the universities. Phys. Soc. 15(3/4): 5.
25. DUNCAN, C. 1986. Strategic computing. *In* Second Annual Report of the Defense Advanced Research Projects Agency. DARPA. Washington, DC.
26. GOLLON, P. J. 1986. SDI funds costly for scientists. Bull. At. Sci. January: 26.
27. SANGER, D. E. 1985. Campuses' role in arms debated as "Star Wars" funds are sought. N.Y. Times. July 22: A1, A12; KISTIAKOWSKI, V. 1985. Science **228**: 304.
28. MYERS, W. 1986. The Star Wars software debate. Bull At. Sci. February.
29. TIRMAN, J. 1984. Militarization of High Technology: 217.
30. SWEET, W. 1986. APS groups discuss SDI impact in physics. Phys. Today. June: 84.
31. KUMAR, C., N. PATAL & N. BLOEMBERGEN. 1987. Strategic defense and directed-energy weapons. Sci. Am. **256**(3): 31-37.
32. WALLER, D. C. & J. BRUCE. 1987. Holes in the impenetrable shield. Bull. At. Sci. October: 5-6.
33. KREPAN, M. 1987. The surprise defense initiative. Bull. At. Sci. July/August: 9.
34. Phys. Today. 1986. July: 48.
35. YUDKEN & SIMONS. 1987. CPSR Newsl. 5(3): 6.
36. Nuclear Regulatory Commission. 1987. NRC News Report **37**(7): 13.
37. ALLARDT, E. 1987. Evolution as a tool for the legitimation of R&D policies. *In* Evaluation of Research: Nordic Experiences. FPR Publication 5: 303-315. Nordic Science Policy Council. Copenhagen.

Index of Contributors

Brown, P. S., 136–148
Bugliarello, G., xv–xvi

Chalk, R., 61–74
Coates, J. F., 149–153
Cohen, D., 184–192
Cole, L. A., 154–163

Dinegar, R. H., 10–20
Durbin, P. T., 47–50

Elzinga, A., 262–275

Feldman, J., 231–244

Hehir, J. B., 99–105
Hollander, R. D., 94–98

Johnson, D. G., 106–114

Kemp, K. W., 115–121

Lackey, D. P., 122–130
Lee, S., 254–261
Liebman, S. A., 164–171

MacLean, D., 34–39
Martino, J. P., 172–183
Michalos, A. C., 216–230
Mitcham, C., ix–xiv, 1–9

Relyea, H. C., 200–210
Roland, A., 51–60
Roth, B., 21–33
Roy, R., 75–85

Schmitt, H. H., 245–253
Shinn, R. L., 40–46
Shrader-Frechette, K. S., 86–93
Siekevitz, P., ix–xiv
Smith, C. A., 131–135

Unger, S. H., 211–215

Weil, V., 193–199